WILDER

In *Wilderlands*, archaeologist Elois
of our changing landscape, through prehistory, Roman occupation, the Middle Ages and beyond. We see the unfamiliar beasts of our old wild make way for species such as brown hare and fallow deer, now romanticised as eternal symbols of the British countryside, but introduced much later than we might think.

Places free from our influence haven't existed for a very long time. But Eloise Kane invites us to rethink our definition of the wild – not as separate from us. Seen anew as the result of millions of human lives lived, *Wilderlands* demonstrates how we are integral to the ecology and biodiversity of our land – with the power to shape its future.

Eloise Kane is a historical archaeologist specialising in the relationship between landscapes, people, and animals. Her PhD examined the archaeology of hunting landscapes, funded by the AHRC. She is Visiting Fellow at the Royal Agricultural University, and Associate Fellow of the Royal Historical Society.

WILDERLANDS

THE HUMAN HISTORY OF WILD BRITAIN

ELOISE KANE

faber

First published in 2026
by Faber & Faber Ltd
The Bindery, 51 Hatton Garden
London EC1N 8HN

Typeset by Ian Bahrami
Printed and bound by CPI Group (UK) Ltd, Croydon, CR0 4YY

A CIP record for this book
is available from the British Library

ISBN 978-0-571-38914-8

Printed and bound in the UK on FSC® certified paper in line with our continuing commitment to ethical business practices, sustainability and the environment.
For further information see faber.co.uk/environmental-policy

Our authorised representative in the EU for product safety is
Easy Access System Europe, Mustamäe tee 50, 10621 Tallinn, Estonia
gpsr.requests@easproject.com

2 4 6 8 10 9 7 5 3 1

CONTENTS

INTRODUCTION
THE BEGINNING

Britain's last wild

Close your eyes and imagine you are standing in the last wild place in Britain. What does that look like to you? Are you with red squirrels in Ballochbuie Forest, the last remnant of the ancient pine forest of Caledon? Or is it Wistman's Wood, tucked into a valley on Dartmoor, a place steeped in legend and filled with stunted moss-covered oaks? Perhaps you can hear waves crashing onto the rocks of the Cornish coast from a windblown heath, where loose ponies graze and gorse grows thick? Or do you stand among the blanket bog and wisps in Ireland, one step away from another realm?

Wherever you are, your mind is likely to have taken you to a place where there are no other people, and no noise from roads or building sites. There is probably a lot of *nature*, of living stuff that isn't human. After all, Britain's ancient wild was made up of the animals we didn't keep – the elk, wolves and boar – and the trees, streams and grass we didn't manage. It was unrestricted by anthropogenic features. There were no hedgerows to protect fields of crops, and large herds could move seasonally as and when they required, unimpeded by us. So if that is what we consider to be wilderness, when *was* Britain last truly wild? And what, if anything, remains?

That question has driven the writing of this book, and pursuit of the answer took me back to the very end of the last Ice Age in Britain, 12,000 years ago. At every site I worked on as an archaeologist, I chased the evidence deeper into the past, recording and attempting to make sense of the strata to understand the why and how of the present. In hunting for remnants of our last wild, the whole of Britain became my site. As with the digging of an archaeological trench, it was possible to peel back the layers and reveal the stories held within different time periods, but the real challenge lay in examining the connections between pieces of evidence to interpret what it all meant. As my interest intensified, I uncovered surprising artefacts that revealed the true scale of the breadth, depth and diversity of our relationships to wild things over time. While we have romantic ideas about what makes up our wild – the brown hare springing from her form or ancient woodlands of gnarled veteran trees – the reality is a far knottier mess, filled with conflicting ideals and beliefs misled by wishful thinking.

When I began my search for wild Britain, it soon became clear it had ceased to exist earlier than most people might think. Our understanding of what it means to be wild is that it sits apart and is separate from us: a wilderness is somewhere unsullied by people. The word 'wild' has always stood in contrast to the domesticated and the cultured: in Old English it was used in *wildeor*, a term for animals that were not under the control of humans; earlier still, there seem to be linguistic connections across northern Europe that conceptualised the wild as something savage and dangerous; another root is sometimes thought to have come from 'will', that the wild is something self-willed.

In Britain, the truth is that places free of our influence no longer exist outside of our imaginations and haven't done so for a very long time. In attempting to excavate a lost wild, another vital story emerged.

Hidden in plain sight

Today, in an era haunted by ecological crisis, ideas of wild things and wilderness take up space in our debates about wellness, conservation, countryside and politics, as well as the stories we tell about our land. The wild has inspired a whole language of its own, a whimsical vocabulary I'll call *wildspeak* – but do we really know what we're looking at or striving for? Wildspeak conjures the wild in our minds as a kind of lost Eden, but in doing so obscures the complexities of real-world ecologies and our entanglements with them. In wildspeak, the moors are bleak, untamed beauty; woodlands are relics of a time before us; animals like beavers become saviours of disrupted harmony. We are told that, in the old days, we must have existed in balance with nature, taking only what we needed from the plants and animals and allowing natural processes to flow around us, uninterrupted. Many of us hold a notion that there remain places where we can reconnect with this ancient way of being, one that frees us from the pressures of modern living. But the nature we see every day has been forced to adapt to a world of our making – and it always has.

Our landscape is an optical illusion: we have been taught to see something that isn't there. When we look out over the countryside we see it through a filter, and our memories conveniently remove the roads, chimneys and all the straight lines. We might satisfy our desire to connect with the natural world by browsing the lush grounds of stately homes, and our longing for the wilderness with a trip to queue at the mountain summit with hundreds of other people. These places are heavily managed, many of them owing their character to their manipulation for agriculture, hunting or leisure. Their very nature is derived from human intervention of some form. We view the landscape as a series of zones and those parts we believe to be the wildest are often considered to have the most value. We have instilled a sense of 'us' and 'them', where everything has its place; carving out space where things are allowed pulls more sharply into focus the areas where we come to believe they are *not*.

Our conceptual separation from the wild has developed over thousands of years, and a collective guilt has sprung up as we have recognised the impact we've had, leading us to reimagine wild Britain. In the face of biodiversity and climate crises, grief and guilt have sent us searching for proof that there is wildlife that exists independently of us, and that nature can recover. But it is inevitably more complicated than this. The stories of how places came to be the way they are involve deep-time geological processes, the actions of animals and plants, as well as people. Evidence of people, living and working and telling their own versions of what they understood to be the way the world worked, is woven through the fabric of our land. It is these traces in the world around us that can be used to explore not necessarily how our wild was lost, but how we transformed it so utterly that we forgot what it was. We eulogise what we think we had but are blind to the bits in between, to all the places where there are hidden groups of beings making their homes in the areas we don't think matter. There is a *wildness* that persists even if wilderness does not.

There is wildlife that thrives with us as well as without us, whether we are looking for it or not. By paying attention, both of these can be found in our everyday surroundings. The primary school in my Wiltshire village is built underneath steep chalk downs on a band of undulating greensand. Here, a small spit of land is flanked on either side by fences, at one end bordered by bushes, and at the other by a path that leads the children from the school gate to their classrooms. Though it is surrounded by activity, the sloping patch of grass between the fences receives very little attention. On early autumn days when the children return to school, as the sun gently warms the greensand beneath the grass, thousands of ground bees emerge from tiny holes and swarm busily. They visit nearby ivy flowers and return laden with pollen, some of the last they will see that season. The bees thrive here because of a lack of human interest in the little piece of land that makes up their home, and they are largely ignored because they are not in the way. In a garden, these immensely important pollinators might be considered a pest, damaging the short-grassed lawn that is such an enduring part of our garden aesthetic. If the fences were to be removed, and the grass patch became an area where children gathered to play, the bees would disappear to another overlooked corner of the village to make their home. This is life in the gaps between what we humans think of as useful space. Tiny pieces of a wild making its home in our world.

In our garden on the other side of the village, ducks are busy overturning leaves and drilling the ground for their favourite slugs and bugs. We keep them for their eggs but they are also useful pest control around the vegetable patch. To protect them from badgers and foxes there is a high fence around their pen, into which they are shooed in the late afternoon. Wiry hairs left on the nearby perimeter fence tell us that badgers sometimes come calling, and we lost a previous flock to an overnight fox attack before we had an electric fence. The ducks' food, a mix of wheat and vitamins compressed into pellets, is kept in a treadle feeder to deter both rats and jackdaws. A bath dug into the ground has a ramp to allow hedgehogs and other small animals to escape, and it also attracts the occasional amphibian. At dusk, the ducks are encouraged into their house for the night to protect them from nocturnal visitors. Keeping ducks has created a set of relationships centred on this small patch of garden under the beech trees. Foxes, badgers, rats, birds, insects, slugs, toads and more are drawn together by our decision to raise ducks. Here, the animals come *because* of human activity. These seemingly small or incidental decisions to fence or to rear this or that impact the living world around us. There is so much happening in the spaces in between.

Archaeological answers

Trips out into nature are often accompanied by the sentiment that we should 'take only photos, leave only footprints'. If we'd ever truly left no trace, archaeologists would be swiftly out of business. We are most often thought of as handling rare artefacts in muddy trenches, but we spend a great deal of time looking at samey bits of pot, rooting through paper archives or recording vegetation and topography at the scale of landscapes, using aerial and satellite imagery to do so. We've even been able to record footprints left by wanderers in tidal mudflats thousands of years ago. As science has advanced, so too have the techniques available to us. We can deduce what type of liquid a vessel held, or where someone who was buried here is likely to have been born. It is possible to observe changes in the shape of domesticated animals through the years, and chart the importance of various food crops. We obtain this evidence so that we can weave it into stories about how people in the past lived and died, and surmise on the meanings they gave to the things around them. Just like the past, archaeology is dynamic, and sometimes our ideas change as new evidence is presented. That we have the opportunity to support, refine or even sometimes upend long-held assertions about our past is undoubtedly one of the most thrilling aspects of being an archaeologist.

Looking at the evidence spread over 12,000 years reveals that the wild of today is not the wild of the medieval period, the Roman period or the Bronze Age. Each has a distinct character, and the people of each period formed different attitudes towards it. But for all of them it was their own *normal.* This is known as 'shifting baseline syndrome' – the baseline is what we accept as normal, and as ecological conditions change, whether through human intervention or not, what is normal changes with them. Once upon a time, it was normal to see spoonbills on the Fens; now it's normal to see parakeets in Hyde Park. Long ago we feared animal predators in the dark outside the walls of our homes; now the only thing we have to fear is each other.

What follows is my version of the story of this shifting baseline that has taken us from an understanding of the wild as part of our world, through to its taming, exoticising, controlling, protection and reinvention. I will argue that there really is no wilderness left in Britain – although I will also suggest that the wild we have is just as worthy as the one we mourn, and the next step in the journey needs to be a rediscovery of it in its own right, and a reweaving of ourselves into what we think of as the wild. This is a tale unique to the islands of Britain and Ireland. Their political and social history is long and complex, and while I often use the term *Britain* for ease, I do so to refer broadly to the archipelago – encompassing England, Scotland, Wales, Ireland and Northern Ireland. These places have never been a homogenous whole, in terms of either landscape or culture, but they are closely enough connected that a shared story can meaningfully be traced across their range. While there are cultural connections between communities here and those elsewhere in Europe, the themes explored in this book are specific and cannot be applied to the continent as a whole. The way in which the old wild slowly disappeared here isn't the same for other regions, some of which today in many ways resemble our ancient past. The spatial limitations of our islands and the density of our population meant that many animals were extinct here earlier than elsewhere, where in some cases, in regions containing larger areas of habitat and less intense conflict with humans, they never died out. In contrast, animals and plants that were brought here were able to colonise quickly because of this density of population and good transport links.

The journey is broadly chronological from chapter to chapter, if not within the chapters themselves. Inevitably, the way we have ordered time into 'ages' in the past means that there are cultural traits that bleed between them. There is very little in history that is as clear-cut as the way we define time, but each block does follow a broad trend in terms of the evidence for perspectives of the wild. It wasn't always the same in every part of Britain, but in each chapter I have teased out exemplars of specific ideas, drawing on places from around Britain as well as on the expertise of scholars from a range of disciplines. At times, I have also drawn on the things I know best – personal stories from excavations I've worked on and places I've been, and on the formation of my ideas from a life that has been rooted in chalk, open grassland and medieval forest.

Humans have biologically, ecologically and irreparably altered the British landscape and everything that dwells within it. While it's sometimes painted as a relatively recent phenomenon, its roots stretch back around 11,500 years, to when humans returned after the last Ice Age and became a permanent presence. Deep down we all know this, but we have romanticised the past and its conditions into a fiction. I wouldn't try to deny we've had a detrimental impact on the environment, but I do think that tackling the ecological challenges we are faced with has to begin with an understanding of how we got here and what underpins our sometimes slightly strange attitudes to what we believe is authentic and valuable. We need to confront the processes responsible, move beyond ecological grief and recognise that we are part of a wild that cannot be separated from us. There is a place for initiatives like rewilding, but only if we can grasp what it really means. While a place's history has sometimes provided the conditions for exciting collections of animals and plants to thrive, in others we cannot always just abandon the place for nature to 'recover'. It would be like unleashing a pack of chihuahuas and expecting them to behave like wolves, or trying to plant a table made of oak. Rather than rewilding everything, I will argue for an ideological reintegration of humans into the wild of now – we must accept we are too entwined to exist independently – and a rekindling of some of the ways we related to the world in the deep past. The wild of old was not something to be used, fixed or appreciated for its aesthetics but was a living and dynamic entity that we depended upon and were part of. By breaking down the barriers that have been constructed and that separate us from nature, maybe we can begin to move towards a future that doesn't have to choose between survival of one at the expense of the other. It is time to see our land as it really is – the malleable product of everything that has come before – and to finally accept that the future is ours to make. What we all do next matters.

1

BEING WITH THE WILD

(MESOLITHIC, *C.*10,000–4000 BC)

The wild taking shape

For thousands of years, a flat and rather inconspicuous landscape in the north of what is now England hid remarkable secrets. Around 11,000 years ago, a craftsperson sat at the edge of a shallow lake with the head of a stag, carefully fashioning it into something extraordinary. The crafting was a messy business, requiring the defleshing of the skull and the removal of the lower jaw, tongue and eyes. Several of the antler tines – the points – were removed and discarded, or perhaps set aside for working into tools for other uses. The remaining antlers were carefully scored along their length and split with sharp flint blades. Much of the skull was packed in clay and placed into a fire to weaken the parts that were to be discarded. It was then broken down so that only the uppermost portion surrounding the antler bases remained, before the thickest parts of what was left were thinned and smoothed. This meticulous work was undertaken to make the skull and its antlers lighter and more comfortable, because this astonishing artefact was destined to be worn. In all, over thirty such antler headdresses have been unearthed over the past eighty years at the Mesolithic site of Star Carr, North Yorkshire.[1]

We cannot claim to know a previous culture's entire worldview simply from the artefacts we bring to light, but we can glean insights into how they went about in the world and interacted with it. All the tools, jewellery, foodstuffs and plant and animal remains that we find provide us with clues. The Star Carr headdresses were certainly about more than boiling down a skull to make a mask. There are two enduring interpretations of the headdresses: one is that they were worn as simple disguises for hunting red deer, to gain proximity to the prey; the other, more intriguing, is that they were associated with shamanic practice, a precursor to, or complementary part of, the hunt. There is good evidence elsewhere in the world for antlers forming part of shamanic costumes, where they were worn and used to communicate with spirits to ensure not just a successful hunt but also continued deer fertility so that populations were kept in balance. That humans believed they could influence the fertility of the deer, through communing with spirit worlds, is emblematic of a mindset that already set us apart from the wild in some way, even when daily life was deeply rooted in and reliant on it.

The deep past, before we farmed and lived in villages and towns, is often presented in popular culture as a time when people were cave-dwellers, hunting and scavenging alongside mammoths and sabre-toothed cats in a landscape that had been unchanged for millennia. This is, of course, not just an oversimplification but largely fiction, a picture of a past so wild that survival was a constant fight. Perhaps the image endures because the remains of such life, so far removed from us in time, are not easily visible as we go about our day in the present, unlike the prehistoric hillforts peppering the uplands or the medieval churches that once formed the beating heart of many villages. In reality, the world of giant deer and cave lions was already in the deep past for the people of Star Carr. By the time their community had developed, those types of iconic megafauna had long since disappeared as the climate had changed. Environmental snapshots from palaeobotanical samples and radiocarbon dates reveal climatic swings that led to the loss of the woolly rhinoceros and the spotted hyaena around 27,000 years ago, followed by the cave lion and the giant deer during a period of rapid change 14,000–11,500 years ago. For some species, like the woolly mammoth, the end was accelerated by humans hunting them for meat, fur and tusks.

Britain was once physically connected to the rest of Europe, only gradually becoming severed as temperatures rose after the end of the last glacial period, or Ice Age, 12,000 years ago. Humans did not continuously inhabit the land prior to that point. Although modern humans arrived here around 40,000 years ago, their presence was intermittent for thousands of years. As the world warmed and sea levels rose, different plants and animals were better able to flourish and so we returned with them. But we did not dwell only in caves or rely on a diet of woolly mammoth, which had already disappeared. The first period after the last Ice Age is known as the Mesolithic ('middle stone age') and lasted from around 10,000 to 4000 BC. During the freeze, the parts of Britain that were not under ice sheets were tundra and grassland, inhabited by arctic foxes and reindeer, but with warmth came trees. While the reindeer left for cooler climes, the elk wandered into the new woods. Britain was largely covered in what we now call wildwood, a forest of natural growth, uncultivated.

This forest is where we first lived upon our return, alongside an old wild which no longer exists because almost all of these species – aurochs (the massive ancestor of the modern cow), elk, wolves, bears, lynx, beaver and red deer – have disappeared. But while the climate largely stabilised 8,000 years ago, the environment was not static. Even during this early period of our recolonisation of Britain we understood that we could have a role in shaping the earth, animals and plants into something of greater use to us. We lingered in places for longer, nibbling at the edges of woodland clearings and burning the upland forests to create embryonic moors. The antler headdresses reveal something of how we lived alongside wildlife but understood ourselves as having the capacity to influence its character for our own gain. So, while the Mesolithic may not have left us hillforts, churches or villas, we can trace everything around us down through centuries of decisions, to the first people to come here and never leave.

The wildwood

What did the wild world look like, the one in which our ancestors set in motion the process which created everything we have today? We can look to its traces for the answer. There is a forgotten forest in Porlock,

Somerset, in South West England, hidden in plain sight. But there is no birdsong here, no fallen leaves or stray sticks, no sweet chestnut casings or conkers – for the trees that stood here have long ago been dissolved into stumps. It is light and open, with no protection from the spray of the sea. This ghost from the past periodically reveals itself at the behest of the moon, exposed, like many others, by the retreat of the tide. It is hard to imagine this place as a living ecosystem although there are some clues: hazel shells lodged in strips of clay suggest there were stands of hazel as well as the larger trees that are present, perhaps oak. These trees were once part of an inland forest that may have stretched over all of what is now the Bristol Channel, from the north Somerset and Devon coast, into Avonmouth and up to South Wales, and which was gradually drowned around 7,000 years ago. The dead stumps of woodland past feature on a surprising number of our beaches, sometimes plain to see at low tide and others temporarily surfacing from their concealment after storm waves have lashed the shore, only to be swallowed back into the sand after mere hours.

The folk memories of these forests sometimes live on in names: the Cornish name for the coastal island of St Michael's Mount, Karrek Loos yn Koos, supposedly refers to a 'grey rock in the wood'. In Wales the submerged forests of the coast are often linked to the legend of Cantre'r Gwaelod, a Welsh Atlantis that was drowned by the sea after a gatekeeper neglected to close the sluices. Generally called 'petrified forests', the wood is not actually fossilised but has been preserved in the oxygen-deprived conditions within peat and clay sediments under water. The stumps tend to suggest either a fairly rapid rise in the sea level or the breakdown of a natural barrier, which caused the inundation of dry land. Events like these would have been significant to people living in the locale, and it may be that some folk tales have authentic roots in the submerging of familiar landscapes. Britain's oldest intertidal deposits, layers of sediments sometimes revealed by the falling tide, are over 100,000 years old, while the youngest drowned forests date only to the last century. Many of them, like the trees of Porlock Bay, are Mesolithic, a fleeting visual record of the wildwood that once was.

The wildwood of our imaginations seems often to appear like the most majestic of our modern forests – dappled light darting between the dense crowns of huge, knurled oak and beech trees, mossy glades peppered with toadstools, silent but for birdsong and wind. This idea of a dark wildwood of giants and ferns is incredibly powerful and appears as a theme again and again throughout the story of our relationship to wild things, often as a theatre of the unknown and the dangerous, in tales like *Beowulf*, 'Red Riding Hood' and 'Hansel and Gretel'. It's true that Britain came to be covered in extensive woodland but it didn't always look like the stuff of castles and gingerbread houses. The reality is our early post-glacial landscape held little of the stateliness we tend to attribute to the primeval woodland we imagine was here for so long. It was rather thinner. The people who first followed the retreat of the ice found landscapes of clumps of narrow birch trees and feather-headed reeds in swampy mires, tall herbs and occasional willow.[2] The environment was more changeable than we think, too, undergoing a major transformation between the beginning and the end of the Mesolithic.

Encompassing the entirety of the island of Ireland, Scotland and almost all of Wales, Britain's Pleistocene ice sheet cut north-east/south-west across England, roughly from the Bristol area to Lincolnshire. Kilometres thick, its shape and size varied as it waxed and waned asymmetrically, so that the ice was never at its maximum reach on all sides at the same time. Powerful glaciers carved huge valleys like the glens of Scotland. Elsewhere, boulders and debris were eased into place where they are still visible today, in rocky landscapes like County Cork, Ireland. The ice did not melt suddenly but retreated over a period of millennia, starting 26,000 years ago. It took 8,000 years to melt northwards so that only half of Ireland, Scotland and the highest parts of Wales and England were covered, and a further 6,000 years to release the north-west of Scotland in around 12,000 BC.

By this time Ireland had become an island, detached from the landmass of Europe. But for this, Britain remained connected by land to continental Europe in the east, across what is known as Doggerland, allowing animals and people to cross during the burgeoning warmth until this, too, was submerged 8,000 years ago.

Britain's tundra south of the ice was a treeless and open environment dominated by hardy shrubs, mosses and lichen. Large mammals like horses and reindeer were able to roam far and wide. As the temperature increased, new species were able to flourish in a land that had become ripe for colonisation. Birch blew in on milder winds, along with Scots pine and rowan, establishing the early woodlands of the Mesolithic. Eventually, these were to be pushed northwards to become the Caledonian Forest of Scotland, and hazel spread into the gaps, followed by alder and eventually the slow creep of elm, oak and lime, over a period of several thousand years.

We are able to illuminate this broad sequence through the collection of palaeoecological samples, sometimes taken during archaeological research but also a discipline in its own right. Cores of sediments reaching metres into the earth can be extracted and analysed for the type of evidence that tells us what past environments looked like. In the right conditions we can find leaves, twigs and seeds that tell us directly what plants were present in the area or perhaps brought to the site. Insect remains can also add crucial information as these tiny creatures tend to inhabit quite niche environments: for example, the oak pinhole borer beetle will only be found on dying or dead wood and can indicate mature woodland, while the eccentric vallonia snail thrives on grassland. Remarkably, we are also able to identify plants from their pollen rain, the name given to the clouds of particles that waft silently to the earth and become sealed beneath layers of other plant detritus. However, while these techniques can be powerful, they are not a magic bullet. Sometimes the pictures we build up of an old environment are relevant to only the very smallest area. Pollen is able to drift, and some insect-pollinated species of tree barely produce any at all. Some grassland-loving snails, too, have been shown to live in very small areas, only metres from woodland. Where wood is identified from charcoal in a hearth, it indicates only that that specific wood was burned on a fire, not necessarily that it was growing nearby. New data emerges constantly that helps us to refine and reinterpret what the world was like; having one eye on critical evaluation and being open to these shifts is an essential part of doing archaeology.

It is one thing to understand what type of species were present, but another to work out what this wildwood may have looked like to the people living in it. Whether or not, once settled, it had a dense canopy or was broken up with naturally-occurring clearings is the subject of long-standing debate. The traditionally accepted narrative was that Europe had been covered in a dense, dark, primeval forest of oak and hazel, with few clearings, and that all interruptions in woodland cover were indicators of deliberate human action. At the turn of the twenty-first century, however, biologist Frans Vera suggested that hazel and oak could only have proliferated in a landscape that was more open, as the species are shade intolerant. His argument was that this was achieved by the natural action of large herbivores whose grazing and trampling maintained clearings, providing hazel and oak saplings with the light they required. This would have appeared more like a wood pasture, with some dense groves, as well as patches of open grassland and regenerating scrubland. It is a popular idea, but also not without its critics. Because Ireland had become isolated from the landmass of Europe during the last glacial period, there were no large mammals – no large deer, elk or aurochs – present on the island at the beginning of the Mesolithic. But even so, the pollen evidence suggests that the island was still dominated by oak and hazel.[3]

I suspect that now is the time to break the news that there is no wildwood remaining in Britain. The last of it was slashed and burned and cut down centuries ago to be used in house- and boat-building, keeping the hearth alight, making tools and furniture, or clearing the way for more productive imported species or parks and grazing. We have woodland that approximates that which may have been found in the past, in small patches, but no coherent habitats that might function in the same way, with the same species. That interplay between red deer, people, hazel, oak and a lake drying out at Star Carr no longer exists anywhere. Even our 'ancient' woodland is defined as anything that has been wooded since 1600, and while these trees would have been saplings during the last years of Elizabeth I's reign, 400 years is generally not considered *ancient*. But there is power in this word that seems to give us comfort and invoke a sense of continuity from the deep past to the present, to connect us to things we yearn to experience again. It is through the lens of the word 'ancient' that we often see our countryside, and it *is* an ancient creation, but one that has been changing rather than static for thousands of years, whether due to human activity or climatic variation. Even during the lifetimes of the people at Star Carr there were dramatic climatic changes that saw temperatures cool, in one event down by 3°C in as little as a decade. This was enough to halt the rise in birch woodland and saw increases in the types of shrub, like juniper and willow, that grow in more open ground.[4]

Not only is our wildwood gone, but in almost every European-wide study Britain commonly features as one of the very few areas in Europe that has absolutely nothing that is even considered primary forest, meaning a forest free from human involvement – one that has been continuously wooded and where natural processes can operate dynamically without impairment. Not one of our woods has been left alone for long enough to act normally again; of the two beech woods that one study identifies as *potentially* primary, both are part of larger amenity woodlands in tourist areas.[5] The UK, in particular, spends the highest proportion of its budget for forests and woods on the category of leisure and amenity, and it also spends more money on this than any other country in Europe. The priority here is ensuring that these are – and remain – human spaces above all else. They have become a spectacle.

As disappointing as this is, the poor state of forests in Britain is not entirely unique. Across Europe, less than 4 per cent of forest cover is considered 'primary' and not all types of forest have surviving patches that fall into that category. The Atlantic region, of which Britain is a part, has no examples of primary forests of particular oaks, for example. What this means is that for some types of forest we simply don't know how they would operate without human influence, or what really constitutes their wild state. A forest is more than the sum of its parts: it is a melding of all of the insects and microbial beings and mammals and plants that make up the big picture. For those types of forest that are extinct in their natural form we currently have the equivalent of a jumble of Lego parts and no real idea of how they fit together or even whether all of them are actually part of the same set. The various components might combine to make many different versions of a whole, but we don't really know which one is canon. The wild forest is so much more than a group of old trees.

Europe's last wilderness

Even with the knowledge that there was nothing left to find, I was curious about whether there was really *anywhere* left that might contain something of Mesolithic Britain, somewhere we can get a sense of that wild before. My search eventually ended with a stretch of forest straddling the border between Poland and Belarus, a place that is hailed as a last surviving relic of Europe's prehistoric forests: Białowieża.

Białowieża is alleged to have been continuously wooded for 12,000 years and is home to wolves, bison, elk, white storks, golden eagles and many plants and insects that cannot be found elsewhere in the region. No single type of tree dominates, and there is lime, oak, hornbeam, pine and spruce, with bushier species seen in dwarf-birch and swamp willow. The forest is protected on both sides of the Belarus–Poland border and the most highly regarded areas in Poland are accessible only on an organised tour, so as to protect its almost unique character.

For all its wolves and rare woodpeckers, is Białowieża really a good comparator for Mesolithic Britain? Is it really all that *wild*? In 2016 the *Guardian* published a travel piece on the forest that claimed it had 'never been exploited or managed', but you don't need to look that far to see that that's not true. The reason Białowieża has survived human activity is specifically *because* of human intervention. At least 500 years ago the area was designated and protected as a royal hunting ground – the name Białowieża itself even means 'white tower', allegedly taken from the appearance of a hunting lodge built in the forest for the use of Russian tsars. By protecting animals for hunting their numbers can be artificially high because the surrounding landscape is managed for them, upsetting the natural balance. On the other hand, overhunting has the opposite effect: a century ago bison were actually hunted to extinction in the wild, only to be reintroduced from captive populations a few years later. So, while there are now hundreds of free-roaming bison for safari visitors to see, they are not descendent from an unbroken line of wild animals. The bison of Białowieża are the progeny of animals at one time kept captive.

Archaeological research is also now revealing that long before the hunting reserve was created, humans were leaving their mark on the forest. Using airborne laser scanning, known as LiDAR, it is possible to build a relief model of a place by bouncing pulses of light off the ground and measuring how long the light takes to return. Processed data can yield a relief model and show archaeological features in remarkable detail. In wooded areas it is possible to 'see' through the leaf canopy where the light passes through the gaps. Using this technique on Białowieża has already revealed hundreds, if not thousands, of recognisably archaeological sites, including swathes of burial mounds and field boundaries that indicate formerly open space.[6] This is supported by the pollen record, which contains evidence of crop planting from late prehistory, as well as the remains of smelting sites that suggest wood harvesting. Archival records attest to the continuing use of the forest by the people living on its fringes, practising tree beekeeping and scything grass for haymaking. Much like the ice sheet, the forest has been eroding and regenerating over time. There is no doubt that the forest is of enormous importance for biodiversity and nature, but it is certainly not the static stalwart of wilderness that it is sometimes presented as.

The most protected parts of Białowieża, the bits that are operating with the least human intervention, don't contain the clearings that Vera's hypothesis suggests were naturally created and maintained by large herbivores in the prehistoric wildwood. Some say this is because modern population management means there just aren't enough large herbivores to make this happen and that they would appear if the bison were simply left to it. The significance of this forest for our story is that it shows that while we can reconstruct elements of our past environment, knowing some of the animals and plants that were present in particular places, definitively identifying the way it all worked is not yet possible. Perhaps Białowieża is as good as we've got, slightly faded but with a lingering authenticity in its recovered natural processes that serves us flashes of a world that is otherwise out of reach. But it also underscores that the wildwood was never one thing, not the endless and mystical land we often imagine, but changing in character over time as plants and animals took hold or faded away, and as people used it and exploited it.

At some point in time, at some point in space, it is very likely that Britain really did have something resembling Białowieża. When we compare evidence from different sites, there emerges a picture of a range of environments that covered large areas but differed depending on the underlying geologies, topographies and interplay between people and the climate at the time. The early Mesolithic lakeside environment in which people fashioned the antler headdresses at Star Carr featured rushes and reeds, the surrounding land covered in familiar plants such as rosebay willowherb and nettles. The first people to call Star Carr home in around 9,300 BC were surrounded by clumps of silver-trunked birch, bright yellow in autumn, which was able to grow freely and expand alongside similar trees like the long-leaved drooping willow. For their descendants, a few centuries later, the lake was becoming more shallow as peat built up. The drier ground meant plants were rooting so successfully that it was necessary to burn the vegetation in order to keep enough open areas for animals to graze in, to perhaps encourage edible herbs for forage. They fashioned a platform of birch and alder wood, ensuring dry access to the lake shore, and it was this community that created the antler headdresses. As time wore on, people visited and maintained it, despite the challenges posed by an environment that was morphing into something else, driven by climate change. The great-great-grandchildren of the platform builders now built one of their own, so it is clear the lake remained enough of a resource to draw people back time and again. As the lake became drier and drier, at some point after 8,500 BC, activity dwindled and people stopped visiting. Over the next 2,000 years the entire lake disappeared and gave way to boggy fen, the old platforms now encased in peat along with all the evidence of over 700 years of intermittent human occupation. Beyond Star Carr, the rest of Britain was also changing and our old familiar friends the elm, oak and lime had moved in. In high places like the Pennines, the birch clung on to survival in the very highest parts, with the lower slopes covered in hazel and oak. Here, people continued to burn clearings into the trees, the evidence showing up 8,000 years later in cores excavated from ancient environments locked underground as thin, blackened strips of charcoal shot through with peaty layers.

The burning of clearings was really important to Mesolithic people, but also to our story: it is a very early example of deliberate environmental manipulation and control. Often, grassland that is left alone will morph into scrubland and then from young to mature forest in a process called 'succession'. It is possible to halt the process – as Franz Vera suggested, frequent trampling by aurochs and large deer can prevent succession from swallowing a clearing. Burning also plays its part: lightning strikes or extreme weather conditions can cause fires which spread quickly, all too familiar to us in the present as we become accustomed to the increasing frequency of destructive forest fires around the world. We know in many cases that the fires we find evidence of in the archaeological record were often set by people – there are too many coincidences in where they occur, especially in upland places and on wetland fringes like Star Carr, for them not to be caused by humans. Various stages of succession offered different resources. The scrub stage yielded blackthorn, crab apple and raspberry. It could also encourage hazel, which was one of the most important plant types across this period and later prehistory. The nuts store well and provide much-needed fat in the diet, and the wood could be used for tools as well as shelters. After cutting, it regenerates quickly and can be encouraged to grow usefully straight poles, which we still use today for building fences. Fish traps from the Mesolithic coastal site of Tybrind Vig, in Denmark, were made using hazel withies that were so straight and in such high numbers they can *only* have come from a place where people were coppicing the trees.[7]

In other places, like the thin chalky soils in the south of England, there were probably more clearings just because the land couldn't support such dense forests. These would have been easy to maintain, and a reliable place where hunted animals could be found. In the rolling hills that would become home to Stonehenge later on, patchy woodland had clearings that are thought to have shifted over time. Remnants of early colonising pine have been found in pits that predate the stone circle by thousands of years. At the nearby site Blick Mead, sedimentary ancient DNA (sedaDNA) shows there was a higher proportion of willow and aspen than would be expected, which may be down to its riverside location. An open glade near to water, like the wetland edges that were burned at Star Carr, would have been somewhere the hunter could wait for their game, and the animal bone evidence from the site shows that this community were eating aurochs, roe deer and red deer, as well as salmon and pike.[8] This was not the passive occupation of an unchanging environment – Mesolithic communities didn't just scrape an existence and move from place to place, but took an active role in making the environment what they needed it to be.

The power of trees

There is a nostalgia that often accompanies our depictions and discussions of the past, particularly when it comes to forests. It's as if we imagine ourselves to have been guests in a landscape of things that were older and wiser, but I don't know whether or not people in the past understood their surroundings in quite the same terms.

In September 2023, one of Britain's most famous – and favourite – trees was felled in the night. For a few days, the outpouring from the public was more akin to a beloved celebrity passing than one might expect for a tree, but this one was particularly iconic. It stood sentinel atop Hadrian's Wall, nestled in a natural dip formed by meltwater under the ice sheet thousands of years ago, in what had become known as Sycamore Gap. There were calls, and even attempts, to replant it, arguments over whether to let it regenerate or replace it with a more permanent sculpture. There is no doubt that the view of the tree in its gap was strikingly beautiful, but what is it about a tree that causes such emotive outpourings? We have a special relationship with trees in the present. They become focal points, and perhaps particularly impressive trees always have been. We hang things from them, from the intangible to the very real. There are myths and legends attached to trees and love notes carved into bark; they have been hiding places for kings and used as gallows. We choose individual trees to celebrate Christmas and plant whole forests to provide long, strong timber for buildings and ships. The oldest and gnarliest of trees are often attributed names and have a character of their own. Whether it is in the shape of their canopy and the form they make on the horizon, or if they mark a place along a familiar route, they can be instantly recognisable.

My own favourite trees, my version of Sycamore Gap, are a deliberately planted decorative 'clump' that tops a hill looking over the Roman road alongside which I grew up. Seeing the distinctive lop-sided shape of the clump as a young child always meant I was nearly home. I have often wondered over the years what that same view will look like when they are gone – the trees are a dependable part of one of the landscapes and vistas I know best. It takes time for trees to form this kind of view, but once they have grown to a certain size they become fixed in our minds. Although always slowly changing, these trees appear in our minds the same as they always have done. But even though many have an average life expectancy that outlasts that of a person, they do not last forever, and in the winter storms I find my thoughts wandering to that exposed knoll and mulling the potential loss of my beloved trees. When they do come crashing down, that view will never be the same, just like Sycamore Gap will never really be the same. How many of these places that are laden with such a weight of assumption, an expectation that they will simply be there for us forever, have disappeared without a trace? How many landscapes of meaning have been snuffed out by the loss of something so ephemeral as a tree? We sometimes find traces of specific trees in the archaeological record – tree throws where a storm has knocked one down, the pollen rain of a forgotten season, or stumps on the beaches. Maybe some of these places are the Sycamore Gaps of the past, one-time places of pilgrimage and wonder, the loss of which prompted people to move on.

Perhaps this fixation with old trees is why we cling so much to the idea of an unadulterated wildwood of dependable ancients, through which we imagine our Mesolithic ancestors made their way, and upon which they based journeys and tales. The environmental evidence, however, does not paint a sharply focused wildwood of Mesolithic Britain but more a nebulous wild*scape*, a mix of smaller and larger trees that came to settle into localised patterns in line with the climate and influenced by the landforms that had been carved and moulded by the ice sheet. If there ever was a Tolkien-like forest of knurled old oaks, it developed only after people were already living in Britain, as one of a series of different habitat types that morphed as the world changed. In those early forests of thinner, smaller species like birch, the life expectancy of trees was also much shorter than for the majestic oaks that came later. It is unlikely so many successive generations knew specific trees in those earlier days.

There is no doubt that trees were important, not just for building but for other uses too. Pine and birch afforded a variety of materials other than wood, such as bark for containers and sap for both gluing and hafting tools, and archaeologists have even found lumps of birch-bark tar with tooth marks that point to medicinal chewing.[9] But other parts of the landscape – for example, geological features or particular configurations of rivers and valleys – may also have taken on the types of meanings we sometimes attribute to trees because of our absolute reliance on all the things within the landscape to sustain communities through the different seasons. It is very tempting to think that during this period small groups of people moved in who left very little trace, who trod lightly on the land and maintained balance in all things, but life was much richer than that. So much of how we interacted with this land of old is now invisible, but through the careful excavation and collection of archaeological materials we can begin to build pictures of the rhythm of life in the wilder past.

Finding life

The culture of the Mesolithic was one of nomadic or potentially semi-settled hunter-gathering, and we've seen that people were already beginning to affect their surroundings. We know from excavation that meals included aurochs, deer, boar, seals, pollock, limpets and mussels. Plant-based food took the form of hazelnuts, berries, apples, pears and a range of other species we might consider unusual today, including waterlily and the tubers of star-like celandine. We did not rely on cattle corralled into fields or crops planted in rows, nor live in extensive settlements of permanent homes – although we do have evidence for some structures. In Britain there was not yet a culture of ceramic production and so there are no pottery remains for us to find. The archaeological record is more ephemeral than for succeeding periods – it just doesn't have the permanence that we see in the monumental building of later prehistory or the castles of the Normans. The way Mesolithic people affected their surroundings was, as we have seen, more a shaping of environment than of landform. Fewer above-ground clues mean that finding sites can be quite difficult. Evidence from full-scale occupation sites are rare, and many finds have been made by chance: caves blasted open during construction work, flint scatters found after ploughing.

Because there was no pottery or metalworking in early prehistory, most of the remains we might find are organic and in ordinary circumstances rot down over time. One major exception is stone, often shaped from flint nodules and used for a huge variety of tools, from arrowheads to blades and awls for leatherworking. Bone often survives, except in the most acidic soils, and antler can also be preserved. Evidence from plants and wood, as artefacts rather than identified in sedaDNA, is best preserved by the anaerobic conditions under water. Given what we know about sea-level rise, particularly the drowning of our land connections to the rest of Europe happening only 8,000 years ago, it is clear that some of the best evidence for how we lived might actually lie locked under the sea and, sadly, most of it will stay that way. There is one site that is accessible and undergoing regular periods of investigation, however: Bouldnor Cliff, off the north coast of the Isle of Wight in the south of England.

In the 1980s, having been presented with a suite of flint axes that had been found by chance in the nets of an oyster-fishing boat, the county archaeologist for the Isle of Wight took the opportunity to accompany a day's fishing to investigate whether this was a one-off find. It very much turned out *not* to be. A whole range of implements, from bone to stone, poured out of the oyster nets and it emerged that fishing vessels had been finding this kind of stuff – frequently – for over twenty years. So began a relationship between archaeologists and the fishing community, who passed on lumps of peat and other finds for examination, picked out when sorting the oysters. We now know that these were not just submerged stray finds, but that they represented an entire drowned landscape full of evidence of prehistoric life.[10]

For much of the twenty-first century Bouldnor Cliff has been subjected to successive seasons of fieldwork recording. Much of this is done in situ rather than being fully excavated because – with the shift of the tides – what is there one day may not be there the next and excavation is much more time-consuming. Divers descend metres under the water, down a steep cliff face to peat deposits overlying sandy clay filled with tree stumps, roots and other treasures. Lobsters have burrowed into the prehistoric sediments and can occasionally be seen throwing unwanted flint tools from their homes. Among the finds, the site has yielded hazel leaves that look as though they were gathered recently from the forest floor on an autumn's day, and string that is an incredible 8,000 years old, twisted together from a fibrous plant such as nettle or honeysuckle. One of the most exciting recent finds was the discovery of a wooden platform of worked bark and sapwood, exposed for only days and carefully recorded and removed before it could be washed away. Given the shaping and nature of other wood found in the area, it is thought this may have been a platform for boatbuilding. Activity at the site dates to the period immediately before the area was inundated with rising seas, and it would have been a wetland environment next to a forest. As the water table rose, people scouting the edges of ponds and along streams would have come across fallen trees that were easily accessible for shaping. Boats allowed exploitation of all the rich resources offered by this environment, but also a means of transport – even perhaps between Britain and mainland Europe to the south across the Channel. The platform here is rare but not unique: Star Carr had three of its own built of hewn aspen, though their purpose is not known. Perhaps they, too, were for fishing or launching boats on the now-disappeared palaeolake, or maybe they were for the deposition into water of the artefacts that have been found to surround it.

Water was obviously an important resource, but the marine environment gave access to a different kind of wild food than on land. A number of sites, particularly in the west of the UK, from Cornwall, through South Wales and to Scotland, are found on the coast. Reliance on seafood is unlikely to have been unique to this area, but evidence is scarcer elsewhere because rising sea levels drowned so many of the lowlands in which people had made their homes in the east. On the island of Oronsay off the west coast of Scotland, shell middens – mounds primarily consisting of discarded shell material – include a high proportion of limpet shell, some periwinkle, fish and even sea urchin. The fish is a type of pollock, killed at an age that suggests winter occupation. As the middens were built up over centuries, it may be that one or several communities moved along the coast as resources were depleted, returning after the beach and rocks had been left effectively fallow to regenerate, or possibly that each midden was used in a different season.[11] Bone harpoon and barbed points are also found across Britain and northern Europe, indicating widespread fishing, whether from boat or shore.

Seasonal occupation of sites can be inferred by the presence or absence of certain species in the archaeological record, such as the pollock above. A lack of shed antler and charred hazelnut points to the absence of people from a place during late summer and winter, and high levels of eel or salmon suggest an abundance of prey as a result of targeted hunting during migration season. We can also use artefacts to consider the activities that went on in places and how these fit into a seasonal rhythm. A collection dominated by flint scrapers, used for scraping and processing hides, may represent a specialised and temporary hunting camp, where animals were killed and processed on-site before people moved on. A wider range of tools and associated domestic activity might be interpreted as more of a basecamp-style inhabitation, from which groups fanned out to gather resources and then returned. We can therefore imagine that some communities moved around different landscapes with the seasons, from uplands to lowlands to coast, and perhaps followed the migration of large mammals like aurochs, as well as the salmon runs.

This understanding of the world, where to go and at what time, was critical to a lifestyle that was heavily dictated by the availability of wild resources, as opposed to the direct farming and stockpiling that was possible later on. There has been some speculation that people kept time by the moon, having recognised the regularity with which it grows and shrinks, and that this might have been used to inform them when it was time to move on. Just keeping time by the moon, however, would see people drifting off course, because there can be either twelve or thirteen moon cycles in a solar year. At Warren Field, Scotland, it has been shown that a series of twelve pits dug in a slight curve 10,000 years ago aligns with a natural pass through nearby hills, where the midwinter solstice sunrise was visible. This effectively made the whole landscape a calendar by which time could be understood by the lunar month, accounting for the drift – move on to the next pit with the passing of each moon cycle, and start again at the rising of the midwinter solstice sun. By doing this, communities unlocked the future: they would have been able to combine lunar, solar and environmental indicators and have a good idea of what was coming. It stands to reason that a landscape-scale calendar like this required a clear view from the pit alignment to the right point on the horizon, indicating that it was created within an *open* space. Might the community of Warren Field have deliberately created an opening that facilitated not just the practical provision of food but also provided a view? The environmental evidence is complicated, but the pits seem to have been neither in open grassland nor on the woodland edge. This seems to have been a birch and hazel wood with extensive heather for its understory.[12] Certainly it was not an opaque barrier to visibility, especially as it was aligned to the midwinter rather than the summer solstice, when the leaves had already fallen, but it surely required constant upkeep to ensure the pits were not lost to the heather.

Even if diet and other activities were dependent on what was available, it seems that not everyone was actually moving seasonally. House structures have been identified at sites like Howick in Northumberland, on the Isle of Man and at Mount Sandel in Northern Ireland. Some of these use natural depressions, perhaps tree throws, surrounded by postholes that would have supported a roof of some kind. Dating evidence from charred remains found in hearths and in spreads of debris within the structures show that some of them may have been in use for centuries during the eighth millennium BC, even if it was intermittent.[13] The occupation may not have been entirely continuous but does show that these places had something of a permanence to them. Within the refuse of these homes is the suite of wild resources that was exploited – hazel and apple or hawthorn for building with, hazelnut and wild meat for consumption, flint for working wood, meat and hides. People were not just processing their catches here, but living. Activity tended to focus on the hearth, the heart of the home even 10,000 years ago.

You are what you eat

In addition to houses, pits, antlers, plants, string, platforms and stone tools, the people themselves have much to reveal about their lives, from their first meal to their last. Hidden in your bones and your teeth are the chemical hallmarks of the food and drink you have consumed, some of the story of you and your past. The common saying is true: you really are what you eat. But you're also what your mother ate when you were developing *in utero*, and what she ate if you were subsequently breastfed. If you eat animals and animal products, you are also what they ate during their lives, and you are the rain that fell in your home town. This is all down to a very useful category of atoms known as isotopes. The periodic table organises chemical elements by their atomic number: the number of protons in the heart of the atom, from number 1 (Hydrogen) to, currently, 118 (Oganesson). The number of protons (and electrons) in an atom is fixed, but atoms of the same type can have different numbers of neutrons. So, carbon will always have 12 protons but could have either 12, 13 or 14 neutrons, expressed as ^{12}C, ^{13}C and ^{14}C – all carbon, but a bit like different species. These are isotopes. Both ^{12}C and ^{13}C are stable, but ^{14}C is radioactive, meaning it decays over time. This is what makes ^{14}C so famously useful for archaeological dating – it decays at a known rate and can therefore be used to work out the age of whatever it has been sampled from. Stable isotopes, however, have different uses. The isotopes we are most interested in in archaeology are found in bone collagen and tooth enamel, and the most commonly used are carbon and nitrogen isotopes, sometimes oxygen and strontium. Much of your tooth enamel was laid down while you developed as a foetus and even if they did not erupt until you were an adult, the enamel of your wisdom teeth was developed by your mid-teens. Once formed, enamel does not change, so its isotopes bear witness to what someone consumed during childhood and adolescence, or, if a young baby, what the mother consumed. Bone, on the other hand, is dynamic and constantly renewing so tells us what people were consuming towards the end of their lives.

Each isotope has a slightly different role in our archaeological detective process. Carbon and nitrogen are particularly useful for dietary insights. The ratios of carbon isotopes vary depending on the types of plants that are consumed, indicating a diet heavy in marine resources, for example. Nitrogen isotopes are passed up to consumers from the food they eat. This makes them a good indicator of how high up the food chain an animal or human was, and therefore whether they were eating more plant- or animal-based products. This chain, from plant to animal to human, means that breastfed babies show the highest values of all because they are being fed by their mothers. Oxygen isotopes in our bodies are derived from what we drink and, ultimately, rain that has fallen and made its way into our water supply. Strontium is specific to bedrock geology, which can show whether someone has lived in one of a range of places with the same signature. Overall, we can bring these strands together to make inferences about things like whether or not someone lived near the coast in infancy and later moved inland, or spent the last ten years of their life somewhere far away from their place of burial. There are limitations and complications, of course, but once we begin to untangle these analyses a whole world begins to be made visible that was previously thought lost.

We do not actually have huge collections of Mesolithic skeletons, especially not from the types of settlement sites described above. For the most part, they have been recovered from caves, and burial methods are varied, even sometimes within the same site. We know that many cave burials were deliberate, but some remains may have washed in. Sometimes we have only single examples of human bone, showing patterns of wear on the bone surface that indicate they were not buried until long after death. Still others show cut marks – evidence, perhaps, of cannibalism, but more likely the result of a way of dealing with the dead that saw their bodies disarticulated (cut apart) before whatever happened next. There are very few burials that can be considered as 'open air', akin to the kind of cemeteries that we bury our own deceased in today. The idea of home at this time perhaps did not include dedicated places for the dead, or maybe that place was in circulation as amulets or similar, like ashes placed on mantlepieces. Despite the lack of huge numbers of individuals to examine, isotope techniques lend context to our other Mesolithic finds. Diet and subsistence seems to have been just as varied as the vegetation and burial practices. A study of human remains from caves in South Wales showed that in some cases diet was overwhelmingly marine, including some consumption of animals high up the food chain, such as seals.[14] But in other places, even those reasonably close to the sea, diet was much more reliant on land mammals and plants. Analysis of bones dredged from the southern North Sea showed contrasting results: some people's diets relied on inland resources, but at other times on freshwater resources, while only a small number exploited marine and coastal foods. Of the terrestrial animals that people were eating from this area, there is evidence from cut marks on bones dredged up that this included the usual aurochs, deer, elk and also birds, such as different types of grouse.[15]

We know, then, broadly what Britain looked like and what we were eating, that we sometimes gathered plants to make bindings and string, and sap for hafting and other uses. We shaped bone into harpoon points for fishing, and used stone to scoop out limpets, throwing their shells onto vast piles of detritus. We took control of some environments, halting the development of mature woodlands to provide a living buffet from which to pick wood for fires and tools, and berries and other fruit and nuts for food. These are the things we did to survive while we lived alongside other animals that had their own cadence of life, who were also affected by the rising tides and other predators that weren't us. I often think, though, that the most intriguing insights in archaeology are not gleaned from considerations of the what or the how, but the *why*.

Animal magic

The basis of so much of how we go about the world in contemporary Western society is rooted in the Enlightenment thinking of the seventeenth and eighteenth centuries, where the logical and natural journey from the past to the present was one of refinement and improvement. Ideas of ritual and magic were eschewed and became seen as 'other', by the monied and learned at least, who began to favour scientific understanding and technological innovation. Reality was rendered as a series of dichotomies: body vs mind, ritual vs secular, human vs non-human. Much of the past has been, and continues to be, interpreted through this frame of thinking. Archaeology is no stranger to it – it's very easy to be lured into thinking that if something does not fit what we think we know to be 'normal', then it must be 'ritual' and therefore special and unusual in some way. What this does is, it stops us from recognising things that were just a normal part of going about the world in another community in another time. We quite often assume artefacts found in watery places have ritual connotations – and sometimes they do seem to be votive offerings, such as Roman curse tablets – but for some communities the practice may not have held the ceremonial meaning we sometimes think but was just a normal part of what happens to certain objects, just as nowadays we put glass in the recycling bin. It's just part of the process of how these objects were used and discarded.

This doesn't make the unusual ways in which we related to animals in the past suddenly mundane. We can see that the headdresses of Star Carr demonstrate that we were deeply entwined with wild things and that our relationship to them went beyond subsistence, but the activity may have been just another fundamental part of the everyday, like gathering wood for a shelter. To understand how animals formed part of the diurnal flow of life in Mesolithic Britain we can look to comparable artefacts elsewhere in the world. The most popular interpretation of the headdresses is that they were worn, for which there are parallels in the shamanic practices of some Evenki peoples indigenous to Siberia. In the late seventeenth century a Dutch mayor and cartographer named Nicolaes Witsen produced an illustration of an Evenki shaman who was beating a drum and wearing an antler headdress reminiscent of those found at Star Carr. Shamanic costume can vary but often features amulets made of animal remains or in the shape of animals, particularly deer and birds. These can be used as protection or as spirit helpers. The Evenki still live in the Siberian forests, and ethnographic accounts of their relationship to the environment provide much food for thought. Wild animals are often either brought or encouraged into settlements, but the relationship acted out there is not one of predator–prey. The wild animals are expected to show their natural selves in return for hospitality, delivering a teaching of their ways to the humans. Animals such as squirrels, who display only a desire to steal food, are seen as rude and are not welcomed back. The Evenki believe that people, animals and even trees have *peredniki*, a kind of second soul which precedes us as we move around real and metaphysical spaces. Human *peredniki* always take the form of wild animals and can appear to other people in dreams. Dogs are able to sense a *perednik* on an empty road and bark into the silence, warning of the coming of people they have not yet seen.[16] In this way, the *peredniki* are there to help see into the future, but are also thought to link people, animals and plants into a shared set of experiences. Going into the forest for hunting, foraging or other activity makes use of this relationship: the hunter will be successful because everything is working together to make it so.

There are similar finds at other Mesolithic sites in Europe, too, and given that we were connected by land to these places it is not a stretch to believe we shared similar cultures. Animal remains are quite often found alongside human remains in caves or other burials – everything from birds, wolves, bears, boar, deer and even snakes and amphibians. Animal teeth make frequent appearances, perhaps because they are strong but small and easily worked into wearable artefacts. Many burials feature red and roe deer antlers placed underneath the body and, sometimes, the metatarsals (foot bones) of deer are found in the chest area, indicating the deceased has been wrapped in a skin.

One of the richest graves we have is from the burial of a woman and baby boy in what is now Germany, interred over 8,000 years ago. The Bad Dürrenberg shaman, as she is called, named after the town where she was found, was between twenty-five and forty years old when she died. She was buried upright in a pit lined with a basket-like structure and filled with red ochre pigment, the baby of only a few months old buried at her feet. Ochre is a reasonably common find in the Palaeolithic and Mesolithic graves of north-west Europe, from Spain and spreading north-east into Siberia, and has been found at some sites in Britain. Its meaning is unclear, but it may have stood for blood or life, or maybe marked graves to prevent accidental unearthing or to denote a spot for remembrance. Genetic analysis has confirmed that the shaman and baby were related through the maternal line, but they were *not* mother and son. The results suggest that there was a fourth- or fifth-degree relationship between the two, making the shaman something like a cousin once or twice removed, or perhaps a great-grand-aunt. It is possible the shaman and boy were not buried at the same time, but as yet radiocarbon dates are not available to compare. DNA analysis also confirms that the shaman, much like the so-called Cheddar Man found in Somerset, was probably reasonably dark-skinned, dark-haired and blue-eyed, genetically determined traits that seem to have disappeared from populations in north-west Europe only towards the end of the Mesolithic.[17]

Among the grave goods were over fifty frontal teeth from aurochs, red deer, roe deer and bison, as well as the bone from the tongue of a boar. A split roe deer bone in the grave had been used to mix or paint red pigment; a lack of wear suggests it was used for a very short time, perhaps created specifically for the burial rite. Other finds included over sixty fragments of tortoise shell, a hollowed-out crane bone used as a container for tiny flint tools, and roe deer antlers with skull attached. It is not just the wealth of animal remains and the roe deer antlers that have led to the woman being interpreted as a shaman. Analysis of her skeleton has revealed abnormalities in the foramen magnum, the passage through which the spinal cord passes to connect to the brain. The changes are likely to have caused neurological differences, seizures and blackouts, maybe marking the woman out as someone who had a higher ability to connect with other worlds in a trance state, assuming a natural place as mediator between people and other-than-human beings. The shaman's appearance as depicted in reconstructions is striking: roe deer antlers and skull hide draped over her head as a mask, boar tusks framing her face and animal teeth hung like a fringe across her forehead. She has a necklace of tusks and teeth around her neck, and her clothes are fur and feather. She is all the wild, and all the wild is her.

These kinds of beliefs, where animals are active participants in a world and can relate to humans from their own perspective, having impacts beyond simply being prey, are not particularly unusual outside of our own culture. On the north-west coast of North America, the Kwakwaka'wakw (indigenous to the coast along the east of Vancouver Island and the opposite shore) celebrated the first salmon runs with activities designed to ensure that they would return the following year. Salmon were seen to have their own society, with leaders and even houses in the ocean.[18] For the Ojibwa of Canada, improper treatment of an animal or its body in death might bring forth the wrath of its kin, who could cause the hunter to become ill or suffer hunger.[19] In the Araucanía area of Chile, a fox stealing chickens is often given the label of 'thief' – not driven by hunger but considered to intentionally commit the crime as an insult to the chickens' owners,[20] similar to the rudeness experienced by the Evenki people from food-seeking squirrels.

If Star Carr's headdresses were about embodying the red deer, of becoming or borrowing the essence of another being to bring about an effect, how might we begin to interpret other things we find? There are other places with unusual animal remains. One of the Oronsay middens described above included nearly fifty individual bones from human feet and hands, as well as from seal flippers. Seals were certainly eaten, perhaps used for their blubber in winter, but they also represent one of the few marine-going species that could move between the sea and land. Perhaps the burial of these elements together represents a form of veneration for the seals' ability to move both in the sea and on land. If you have ever seen a seal on land, you will know that when its smooth blob-like form is not resting on the sand, it is flip-flopping clumsily in an attempt to get back to the sea. Once submerged, though, it is transformed, becoming all the things it is not when on the shore – fleet, nimble, and a formidable hunter. In a way, humans and seals mirror each other – both a daunting presence in their own habitat, and something rather less elegant in the other's. The bones of human hands and seal flippers look extremely similar – might this be what drove the specific deposition of flippers and hands in the midden? I'm not suggesting there was some kind of seal cult in Mesolithic Scotland, but when we think about all of the ways in which we may have lived alongside animals in the past, it doesn't seem that ridiculous.

Animal tales

These attitudes to animals immerse people within a world where our role is not just one of domination. Animals were more than just the non-player characters that wander the land beyond the hearth, but at the same time the hearth itself was responsible for what already set us apart. Along with the ability to cook, keep warm and transform objects, fire brought us something that modern society couldn't function without: artificial light. Torches and lamps made by burning sticks or animal fat meant that, for the first time, we could deliberately illuminate the inside of dark spaces such as caves. This new-found light introduced extra length to our days, the gift of time that had previously been spent either in darkness or by the glow of moonlight. However, as useful as it was, firelight is fickle. It dances with subtle shifts in airflow, might be snuffed out by rain and can grow and diminish in intensity depending on its fuel. Firelight could not replicate the light of day and didn't provide the opportunity to carry out the same tasks. It has been suggested, therefore, that what firelight really brought us out of the dark, when we harnessed it hundreds of thousands of years ago, was togetherness[21] – time spent socially, telling stories by or around the fire, to transmit knowledge of the way the world worked. Here, animals really came to the fore.

The earliest-known art in the world is located in caves. It consists of paint, smeared clay or etchings, and must have been created and viewed by firelight. It may well be why some Palaeolithic artwork takes the form of animals superimposed over one another, so that flickering light could have rendered a moving image of some kind. Cave art doubtless had a role in storytelling, whether it was simply as a visual medium or, more likely, to exist hand in hand with oral traditions that educated new generations in communities. Most prehistoric art dates to before the Mesolithic – the earliest in the known record is up to 45,000 years old and located in Indonesia. In Britain, the rock art of Creswell Crags, Derbyshire, dates to the very end of the Ice Age, at between 13,000 and 15,000 years old. The carvings here are of red deer, bison and horses, the latter two of which disappeared with rising temperatures and hunting, and some figures that appear to be long-necked birds or maybe bird-headed people. If these are indeed bird-headed people, it shows the possibility of a continuity of world-view between the Palaeolithic and Mesolithic – that the boundary between humans and animals was ambiguous and humans could become or embody animals, like the seals from Oronsay and the deer from Star Carr. Cave art and inscriptions may not have been the only way to tell these stories, with fire bringing them to life.

In 1958 a single artefact was recovered during work in the Lepaanvirta river, in southern Finland, which radiocarbon dating suggests was created in around 6200 BC. It is fashioned from a length of wild reindeer antler, which turns at a right angle at its midpoint and twists along its length much like a natural vine. On one of the faces of the artefact geometric patterns have been cut and scraped to crisscross the surface, and in some parts have created a pattern resembling the feathers of a bird. Along half its length the antler has been smoothed and shaped, and an abrupt curve at the end of this side resembles a head, with two carved depressions thought to depict eyes. Interpretation of the artefact has varied from a snow scraper to a shamanic drumstick, but the most interesting of all was posited more recently by researchers who looked at its capacity for telling stories.[22]

When this Lepaa artefact is held up in front of fire and sunlight, a whole Mesolithic menagerie emerges from a slumber of 8,000 years in shadow form. Depending on the rotation and angle of the head, a bear, a snake, elk and waterfowl can all be brought into being. They can appear very big or very small, fade, move and jump. Because the animals and shapes are difficult to control in flickering light and on different surfaces, it is thought that if indeed the antler was used as a shadow puppet, the artefact itself led the storytelling – whatever emerged through the interplay of light, antler and shadow became the focus of the tale. The idea of this shadow storytelling gives dynamism and personality to people who too often exist to us only as a snapshot of science on the page, as a scatter of stones on a field's surface or a jumble of bones in a cave. It makes them more than just a means to an end, more than a way of telling whether people in the past ate more fish or meat. It is suddenly easy to imagine animated individuals telling others of the things they needed to know. Just as the deep, dark woods crop up in cautionary tales across space and time, so, too, do certain animals – a wolf in sheep's clothing or seals as selkies. These stories may have now taken on a role in our folk memory and become old-fashioned, a bit of whimsy, but once served a very different purpose in helping to situate us in the tangled web of life.

The Lepaa artefact is unusual, and we have no direct parallels from Britain. We do have some examples of portable art, though, particularly geometric patterns that may be intended to resemble scales, fur or feathers. The chevrons and zig-zags can resemble the pattern on vipers, a venomous snake that has the widest distribution among snakes in Europe and Asia. The wild of the Mesolithic was not just there for us to hunt in, but was also far more dangerous than it ever would be again. We have fewer examples of predator than prey in the archaeological record, but we do know that there were brown bears present in Britain, as well as wolves, foxes and wild boar. Knowledge of and protection from these must have been important.

In the end

The people of the early Mesolithic had followed familiar plants and animals as they spread west and north together to recolonise a land as it woke up from its freeze. They brought with them their dogs and knowledge of fishing, hunting and the uses of various plants for making tools and shelters. The world was in a state of flux and these early communities had to adapt over time to oscillations in the climate and, eventually, to a settling that had seen the birch, pine and aspen give way to hazel, oak and elm. But even once this had happened the landscape was not one of a timeless and unchanging naturalistic portrait. Regardless of whether or not the natural order would have seen clearings created and maintained by large herds of grazers, people had started to take the reins in instigating and encouraging this process. By deliberately creating clearings, all-important hazel could flourish, and fire and coppicing yielded the straight poles required for fish traps and other uses. Clearings also encouraged an abundance of fruits. People carved bone and rock, painted with ochre, created intricate flint artefacts and maybe told stories of the lives and dangers of wild things with shadow and light. There were seasonal camps, especially on the coast, possibly proximal to semi-permanent bases that served as places from which to follow a natural rhythm. People in the Mesolithic were not living hand to mouth, not entirely reliant on chance finds. They had deep knowledge of their landscapes, a concept of time, of what to do and when to do it, which ensured the survival of people and culture into what was to come. They made choices about where to be and what to shape. I imagine that it is because humans were wholly reliant on wild things that we sought to control them.

In the years that followed the end of this period, 6,000 years after the last of the ice melted, the wild was set to take on entirely new meaning as the hunter-gathering lifestyle disappeared from Britain. Change was accelerating – no longer driven by natural fluctuations as the climate calmed and people moved on from Star Carr, but by humans developing new ways of living and new beliefs. It would be almost another 6,000 years before we realised we had tipped the balance and were now the ones driving changes in the environment and climate all around us.

The first people that burned the first clearings set in motion the adoption of practices that have echoed over millennia and are today visible in everything from hedgerow apples to heathery moors, coal tips to dairy cattle. They had begun the inexorable march towards what we have come to know as the British countryside.

2

TAMING

(LATER PREHISTORY, 4000 BC–AD 43)

The dawning of domestication

On a summer's day sometime around 3800 BC, a medley of people and animals gathered in an open grassy plain in the south of England. They had come to feast, and one or more of the groups in attendance had been hunting for the occasion. These people brought with them the carcasses of at least seven roe deer, which were butchered on arrival. Some of the bones were split open to get at the nutritious marrow within, before the remains were deposited in a large pit of almost two metres across. At least three other communities attended this feast, not contributing the wild product of a hunt, but instead herding their own livestock across the plain. All of the cattle were female, which, coupled with the presence of young calves, hints that the herd was managed for milk rather than meat. These cattle were killed on site and butchered, but because they yielded much more meat than the small roe deer, only the limbs and skulls were placed in the pit, the larger cuts possibly taken away for processing and later consumption elsewhere. When archaeologists excavated these remains several thousand years later they found that there was little to no gnawing on the bones, and so it is thought that the animals were killed, butchered and discarded into the pit in one short episode.

Although the pit may sound like it contained an unremarkable dump of rubbish after a party, it is unique within British archaeology. Pits themselves are common, containing sometimes deliberate and carefully considered deposits of special things, and sometimes dumps of refuse, but nowhere else do we find a mixing of animal remains in one place that represents such contrasting ends of the wild–domestic spectrum. Isotopic evidence suggests that each group hailed from a different area within a 20km radius of the pit, with cattle that were grazed in the open, and roe deer coming from woodland.[1] This, then, was a variegated environment rather than the grassy expanse that it had become by the time Stonehenge was erected nearby, centuries later. By this point nearly 6,000 years ago, we had entered what is known as the Neolithic, or, perhaps more accurately, the Neolithic had arrived in Britain. The Coneybury Anomaly, as it is known, has recently been interpreted as the remains of a meeting between the new farmers with their managed herds and some of the last hunter-gatherers in Britain. A handing over from one way of life to another.

In the past, it was thought that the transition from hunter-gathering in the wildwood of the Mesolithic to the farming of the Neolithic was probably a gradual one, with indigenous groups slowly adopting new practices as they were introduced to the unfamiliar plants and animals that had been made possible by domestication. The evidence now points to the changes being fairly abrupt[2] and that large numbers of people crossed into south-east England from the continent and settled here. Analysis of ancient DNA (aDNA) shows that there was a wholesale shift in the genetic signature of people living in Britain in the early Neolithic – it is entirely possible our meeting in the forest was between people whose genetic origins were vastly different to each other. Major changes accompanied the arrival of new people: the introduction of domesticated cattle, pigs and sheep, as well as cereal cultivation, pottery and new ways of marking our place in the world, from megalithic tombs to massive earthworks. All of these things heralded significant changes in how we related to and understood wild things, and there was no going back.

This chapter covers a period of around 4,000 years, spanning not one but three distinct time periods: the Neolithic ('New Stone Age') from *c.*4000–2300 BC; the Bronze Age from *c.*2300–800 BC; and the Iron Age from *c.*800 BC–AD 43. In contrast to the Mesolithic before, these phases have left us a veritable cornucopia of evidence, much of which we can see around us still. This was when things began to take on a greater degree of permanence, from our ecological inheritance to the pottery and metalwork we began to use and the monuments we built. Each has a distinct enough character from its predecessor to have been dedicated its own 'Age', and there are variations in the form of settlement, burial, travel, trade and diet. What binds them together but separates them from the time before is the emergence of a new way of life that ceased to rely solely on wild things and instead increasingly designed the world to perpetuate the new order. There was not such a big shift in how wild things were viewed and controlled until the advent of the Roman period in the first century AD.

The story I tell in the following pages doesn't dwell on the exploitation of stone and metals that lent each block of time its name, and is not recounted as a linear narrative, from early to late. Instead, this is a story of the new domesticates that appeared on our islands and the impact these arrivals had on our use, exploitation and perhaps fear of wild resources (although there were some surprising 'wild' introductions down the line too). It is a story of disappearing and becoming. The disappearance of much of the old wood but also of old ways, as a new normal emerged that insisted upon permanence, of fields, buildings and commemoration. As the story peeks at snapshots of sites that disclose ways of being in the world, we will weave through mist-covered marshlands, uncovering secrets lurking in hedgerows and lost roads to nowhere.

New people, new animals

Animals are often an excellent proxy for humans in archaeology because our lives have been so entwined, especially with those on which we have built much of our success as a species. Finding animal remains in archaeological contexts tells us about the people who lived with them, how they were eating or using animals and what their attitudes towards them were. The vast majority of animal remains that are uncovered on any archaeological site that formed after 4000 BC are of the big four domesticates – cattle, pigs and sheep or goats. The latter two are notoriously difficult to tell apart and are rarely distinguished into separate species in archaeological reports. Crucially, none of these animals were domesticated within Britain itself. There is no genetic link, for example, between the wild populations of aurochs that lingered on this collection of islands and the cattle that appeared in the Neolithic, nor does there seem to have been much interbreeding between the two. When these animals arrived, they were already very different to their wilder counterparts, the product of a process that had been under way for 6,000 years elsewhere.

The actual process of domestication is much debated. Traditionally, it has been assumed that humans were entirely responsible, making deliberate choices about what was propagated. More recent thoughts have turned to the way in which animals and plants were not just subjected to the will of people but may have had a role in effectively domesticating themselves. In this scenario, wolves worked out that there was an advantage to hanging around near human settlements, with all their surplus food, and it has recently been argued that in the earliest days of deliberate agriculture in the Near East, some plants growing on the periphery of the newly disturbed ground actually competed with each other for growing space. This meant plants were adapting to anthropogenic environments by improving, for example, their own yield or how quickly they germinated. Rather than being an entirely human responsibility, then, this was more symbiotic – both the plants and the humans benefited in the long term.[3] These models are crucial for the way we think about wild things, because they lend the animals and plants an agency of their own. In this way of thinking, it is easier to envisage humans as *part of* the world, as one node in a much vaster set of relationships, rather than the be-all and end-all of existence. We have undoubtedly had a huge impact on the environment, but centring ourselves as the only agents of change contributes to the fallacy that all we need to do to 'heal' the environment is remove the people. I would argue the way forward is not a question of our presence or absence, but rather thinking more deeply about the implications of our being in the world, and what responsibility we have for the future.

It is thought that this symbiosis also played its part in the creation of the dog from a wolf, two animals that today sit at the extreme ends of the wild–domestic spectrum; the chihuahua, the Labrador and the poodle are all different flavours of the same species. There are several theories of how dogs came to be our best animal friends. One belief is that wolves were drawn to human encampments or settlements by our refuse, scavenging leftovers, with both wolves and humans later learning to tolerate each other. In this model, humans realised the potential in partnership and effectively carved off a group to breed from. Another hypothesis is that we took wolf puppies from their wild litters and kept them as pets, eventually creating a genetically distinct group.

The truth is we're not quite sure how many episodes of domestication occurred in the creation of the dog, but the earliest they appear in the archaeological record is around 15,000 years ago in North America. In Europe, the earliest is Mesolithic, where a dog was found buried with a human and dated to 12,000 BC. It was a puppy, just over six months old, and there was skeletal evidence that it had suffered – but recovered from – a viral illness so severe it could only have survived with human care and could never have been of any use in a working capacity. This was an animal that was not culled or left to die but was cared for more like a pet. In Britain, isotopic analysis of a single dog tooth from a site near to the meeting of the Coneybury communities described above showed that it may have come from as far north as Yorkshire, presumably travelling with its human companion. While dogs fulfilled many roles and were able to help, perhaps, in the acquisition of hunted food, they were not often prey themselves, although it may surprise some to learn that they do seem to have been kept for their fur over 4,000 years ago in North America.

Once our relationship with animals had resulted in the invention of an entirely new type of being, it was only a matter of time until more animals would follow, spreading across the globe alongside people. What came next was the waning of the hunter-gatherer way of life and the acceleration of a lifestyle that relied on what we could grow and herd, rather than find and catch. Hounds aside, the early breeds of sheep, cattle and pigs had emerged into being in the Near East, along with other innovations such as pottery and new stone technologies. These made their way here with people in around 4000 BC, as the Neolithic way of life dispersed throughout Europe, with genetic evidence suggesting it arrived in Britain via a Mediterranean route.

You would probably not recognise these first embryonic breeds today. We did not have any kind of sheep here originally; there is no wild British ancestor, unlike for the pig and cattle. And although sheep had been domesticated, the process was not yet complete and, moulting annually, they hadn't developed the type of coat that lent itself to spinning into yarn. Fiona, a sheep who was rescued in late 2023 from the base of a cliff where she had spent two years eating her way through an acre of grass, resembled an enormous cloud because the coat of most types of modern sheep will grow indefinitely. Most of those we see in fields around the British Isles rely on intervention in the form of shearing to keep them comfortable. The earliest sheep to be domesticated was the Asiatic mouflon, which hailed originally from the Levant and is now extinct. There is a surviving feral descendant of these early sheep in some parts of Central Europe: the slightly strange-looking Mediterranean mouflon, with long legs, short brown coat and huge curved horns. In Britain, we also seem to have some survivors of these early pioneers. The Soay sheep, with unmanaged populations now confined to St Kilda in the Hebrides, in Scotland, is more closely related to the mouflon than to any other breed, including the so-called 'primitive' breeds found in Scandinavia. It is thought the Soays may have made their way to the fringe of north-western

Europe and there found their own refugia, where they lived quietly for several thousand years.[4] It was not for over 2,000 years after the arrival of sheep on our shores that evidence for wool textile production shows up in the archaeological record. The sheep of the Neolithic were herded for their meat and their milk, with contemporary textiles still spun from plant fibres.

Coincidentally, as the sheep grew their wool, the last of the aurochs in Britain seem to have been hunted, having clung on in the wild for two millennia after the dairy cow's arrival. Dairy is a hugely important source of fats, proteins, sugars, vitamins and minerals, and having access to it gave people a massive advantage over those who didn't. It is even thought that it was milk, and not meat, that could have been a primary driver towards keeping managed populations of animals in the first place. Despite this, we know that lactase, which helps the body break down lactose in dairy so that it can be digested, was not present in humans in Britain until after 2300 BC, sometime during the Bronze Age. People were either consuming dairy in very small quantities, *or* consuming large amounts despite its ill effects, *or* they were processing it in a way that meant much of the lactose was removed. Evidence from some ceramic vessels shows us that milk processing was common, and it's thought that perhaps it was processed into cheese, making it more digestible as well as portable.[5]

Milk from cattle was taken specifically from the domesticated cow rather than the wild aurochs, and the aurochs' slow disappearance over the two millennia of the Neolithic may have been down to a combination of factors. We know with certainty that they did not merge into dairy herds, and do not survive as ghosts in the DNA of the cows you see in fields today. Trophy-hunting may have played a part in their demise, and it has even been suggested that extinction was a deliberate choice in order to avoid aurochs competing with our cattle, physically as well as for their grazing. Aurochs survived elsewhere in Europe for thousands more years, beyond even the medieval period, and written records and illustrations show them to have been aggressive and feared. A general decline in their mixed woodland habitat as the snowball of agriculture grew must also have been a factor. One reasonably late survivor was found in a large pit at the excavations at Heathrow. Like the Mesolithic–Neolithic Coneybury pit above, it dates to a transitional age: *c.*2300 BC, as the Neolithic gave way to the Bronze Age. The change between these two periods is not as marked as the one before, but rather is characterised by improvements in technology and intensifying agriculture, rather than a wholesale switch to a completely different way of life. The aurochs skeleton at Heathrow was found with the arrowheads that were presumably used to kill it still embedded. The aurochs wasn't eaten or used in any way. It is possible this late survivor's death was as symbolic as the deer in the pit at Coneybury. Where the deer and cattle at Coneybury may have expressed the celebration of coming together in new ways of living, perhaps the death of the last aurochs represents one last foray into the big game hunting that we had been reliant on for so many thousands of years before.

In among this swirl of changing bodies and diets another new animal was brought to the British Isles. This one wasn't really going to change our diets, but it was about to completely reform the way we negotiated space. While the horse had been present in much deeper prehistory, it does seem to have been absent here in the Mesolithic and the Neolithic, reappearing during the Bronze Age after 2300 BC and becoming a normal and settled part of the landscape by around 1000 BC. Once they were here, horses were immediately put to work, used for transport but also perhaps in warfare in the early days. Artefactual evidence reveals specific types of sword that seem to have been designed for ease of use from horseback, with single-handed grips and long blades that were best wielded from a height. While early wheeled vehicles may have been pulled by cattle at first, they were certainly pulled by horses during the Iron Age, at least from 500 BC. If you have experience of riding horses and are familiar with the bit that goes in the animal's mouth, you'd instantly recognise the prehistoric equivalent because they really haven't changed in thousands of years. The 'snaffle' bit normally consists of two rings that sit either side of the mouth, in the corners, with the mouthpiece itself formed of two jointed pieces of smooth metal that sit on the tongue and across the gum between the canines and the premolars. Despite their use for warfare, Bronze Age horses were not like the large cavalry mounts you'd see outside somewhere like Buckingham Palace now, but were more like the feral ones that can be found roaming the moors. These are the genetic descendants of the early prehistoric horses, though an important point is that they are not quite the same in terms of form or behaviour – the ones we are left with have adapted over centuries to live without people, and are the product of a mixing of different types of horse over many introductions and many years. There is no unbroken line of purity here.

Landscapes of living

Building a society in which we could rely on what we farmed rather than what we foraged and hunted doesn't mean that the wild beyond wasn't still a key component of prehistoric life. We will cover the animals below, but it is worth thinking a little about other ways in which people interacted with their environment, one that had been transformed since the Mesolithic from largely wooded to much more open. Many will be familiar with some of the most famous sites of later prehistory, as well as the interpretations that ground their creation and use in the rhythms of the natural world: Stonehenge, for example, or Ireland's Brú na Bóinne (also known as Newgrange) passage tombs, both oriented to the way the sun tracks across the sky on the solstices. Significant monuments such as barrow cemeteries are often carefully placed in relation to topographical features, along ridges and sometimes in valleys. Often, there is an association with water, such as at a site near the Avebury stone circle, in the south of England.

Silbury Hill is a rather strange site. Travelling along the A4 between Calne and Marlborough, it emerges suddenly into view, right next to the main road. At 40m high and distinctly conical, it is striking and quite obviously not a natural feature in this otherwise open landscape of gently rolling hills. For centuries it was assumed to be a giant prehistoric barrow, probably containing the treasure of a rich king. Over many years and many campaigns, as early as the eighteenth century people dug into its top and sides in an attempt to get to what they thought would be a valuable haul of artefacts in the centre. While they came away empty-handed – even a 1960s BBC campaign fronted by David Attenborough failed to reveal the hill's secrets – these years of intervention left their mark on the hill. In the year 2000, the enigmatic monument began to collapse into the tunnels and voids that had never been filled.

By 2007, a joint venture between engineers and heritage professionals began to record and stabilise the site for the future, and I joined the team that summer to work in the public-facing interpretation team. The collapses had presented a once-in-a-lifetime opportunity. Because the tunnels had never been filled, it meant that archaeologists could enter the hill and record the archaeological layers with modern techniques, collecting all sorts of material that had never been given a second thought before, and attempt to piece together the story of its construction and, ultimately, work out why it had been built. Actually being able to go inside this kind of earthwork was a unique experience, not least because at its very centre, in the layers that made up the earliest and smallest form the hill ever took, the anoxic environment meant that the grass that had been cut and piled to make it was still, after nearly 4,500 years, a striking shade of *green*. It was like looking at a living piece of prehistoric nature.

The ultimate goal of the project was to stabilise the hill for the future and not to actually excavate new material but to reassess what had already been exposed in the past. What we can say is that there doesn't seem to have been a burial inside or underneath Silbury Hill, rich or otherwise – it was not a monument to the dead. The hill started life as a small gravel mound, before it was later built up with turves and then chalk over a period of up to around 150 years. For such a huge monument, one that is normally associated with bodies and burial goods, the make-up of the hill itself was surprisingly clean, despite meticulous sieving. The lack of finds has implications for the interpretation of the monument, as well as for the few artefacts that were recovered. It means that this was not a place in which there was much activity other than the construction of the mound; there is no waste from living here. There really was very little that wasn't organic, but some antler picks – which would have been used to dig and to quarry the chalk – may suggest deliberate deposition rather than loss, perhaps by someone whose time working on its construction had ended, or because the antler tools were so important that they became part of the hill.

More can be gleaned from the size and the location of the hill itself. The summer we worked there is memorable for just how wet it was, months and months of seemingly endless rain that caused yet more collapses mid-project, but also flooded the low land around the mound from where the chalk had been dug. Silbury is part of a complex of Neolithic and Bronze Age monuments, among them the Avebury henge and stone circle and West Kennet Long Barrow. The complex is set in a landscape of springs, and a short distance from Silbury is a place called Swallowhead Spring, a major source of the River Kennet, itself a tributary of the Thames. At the time of the hill's construction, the climate was warmer and wetter, and hydrological analysis has shown that while the springs now rise at Swallowhead, they may once have risen right by the base of Silbury. One interpretation of this place, therefore, is that Silbury was built as a communal exercise in monumentalising the source of a major river system. The fact that the gravels and turves that formed its earliest version seem to have been brought in from the surrounding area, rather than taken from the immediate, could reflect the different communities to which this water had deep meaning. Rather than a monument to the dead, it seems to have been a monument to the living landscape.

Wild icons

The character of this living landscape had, of course, changed significantly from the earlier period, when hunting in the woodland was commonplace. Because new animals dominated the Neolithic diet, their remains also dominate the animal bones in the archaeological record. The wild ones are not invisible, but we do have to work a little harder to find them and try to get at their meaning. Truly wild animals were never again hunted to such an extent as they had been in the Mesolithic, but they were still used. Plants continued to be important – the new crops couldn't replace all of the functions that had been fulfilled by wild harvests – but the contribution of foods such as crab apple and hazelnut was much smaller than it had been. The lifestyle of farmers in the Neolithic was very different to that of those in the Iron Age, with the former likely to have included far more movement – we don't have much settlement evidence available for the early period, though houses are occasionally found. Most of the evidence for the Neolithic comes from funerary and feasting deposits. Although this means many of the sites we have excavated were not quite part of day-to-day activity, they do act as windows through which we can find where the wild things were, and they seem to have taken a slightly nebulous form through the long period that encompasses the shift from using stone to metal. The importance of the wild and its identity shifts across time, with some species drifting in and out of focus – sometimes wild, sometimes domesticated. In the Scottish islands, red deer appear to have existed in the space in between, a managed wild as it were, and the special treatment of wild boar and pigs suggests a symbolic significance of their place either side of the wild/domestic divide. Some of the best evidence we have for exploring attitudes to the prehistoric wild can be found in the archaeology and iconography of birds, held in high regard for thousands of years – but not for their meat.

Avian allegories

On wandering through the European prehistory exhibits at London's British Museum, you may come across a long bronze implement, notched with grooves encircling its shaft and curving 90 degrees at one end to terminate in two sharp, vicious-looking prongs. Uncovered in Ireland in the early nineteenth century, it was, like many of the best things, found in a bog. While unusual in its appearance, this particular type of tool is not unknown and has parallels across the Atlantic seaboard; it is actually a flesh-hook, used for hooking stewed meat out of huge bronze cauldrons during large feasts. It dates to around the eighth century BC, the Later Bronze Age, and is an example of the importance of communal eating at this time. The flesh-hooks were often deposited in wet places when they were no longer to be used, so much so that it's thought this deposition, which it can be tempting to see as a 'throwing away', was actually an integral part of their use. It seems to have been an important last step in their use-journey, a burial into the bog or marsh. The most fascinating thing about this particular flesh-hook is the animals that are dotted along its length, fixed through the shaft of the hook and seemingly designed to move. The way it has been reconstructed today features two avian figures at one end, thought to be ravens or crows, facing five more birds of a different form: two adult swans and their three cygnets.

This kind of iconography is rare but insightful: both corvids and swans have their place in folklore and mythology generally, as well as in the archaeological record in Britain, particularly during the Iron Age in the first millennium BC. In Ireland, corvids are associated with the war goddess Morrigan, who can assume the bird's form, and the Celtic god Lugh fulfilled a similar role elsewhere.[6] Corvids are highly intelligent and sociable birds, capable of mimicking human speech and stockpiling food, and are so tolerant of humans that they can be tamed. They are also scavengers and must have frequented settlements where refuse was present. Keeping domesticated animals at this time meant having a breeding population, providing a semi-reliable source of young carcasses for scavengers. Ever more settlement during the Iron Age can only have served to exacerbate this, and the deliberate burial of corvids, an unusual feature of these last centuries of prehistory, may reflect their tendency to stay near to people. Often found in the bases of pits, a lack of wings on some skeletons suggests that these birds may have been killed for their feathers.

Swans also often feature in stories of shapeshifting, humans taking the bird's form to permeate the boundary between this and the other world. This seems to be linked to species that move seasonally, and some contemporary beliefs in the Baltic region assigning special significant to migratory swans may extend all the way back to the Neolithic – swan engravings from this date in western Russia can be found in association with cracks and fissures in the rock, leading to an 'other' world.[7] We cannot be certain that these oral traditions have roots back into prehistory, but it is significant for corvids and swans to feature on an implement used for feasting, which was about far more than consuming lots of food but was a possible contributor to, or a direct consequence of, domestication, requiring either a surplus and/or significant co-operation or organisation between contributing groups. As at the Coneybury gathering nearly 3,000 years earlier, it may have been a way of cementing alliances or kinship ties. The birds on the flesh-hook must have had some meaning in the context of the consumption of farmed animals, but more broadly wild birds were woven through much of prehistoric life and death.

Tomb of the eagles

Across the world, where birds take on roles in belief systems they are often linked to the dead. Corvids and other birds will eat a human body if it's available, such as those that are exposed through excarnation. This practice, where a body is left out in the open to be broken down by exposure to the elements and other animals, was reasonably common at various points during later prehistory, and persists in other parts of the world today. There are even cases of birds and other animals being buried with the dead. But not all are equal: different animals symbolised different things, and not all had equal importance in terms of their symbolism. On the Orcadian isle of South Ronaldsay, the remains of dozens of people were placed together in a clifftop tomb. Discovered in the 1950s, when the farmer noticed some stones eroding out of the bank, it is one of the most famous animal–human deposits of the Neolithic, and is now known as the Tomb of the Eagles. The structure is a chambered cairn, built of stone slabs in an oval shape. It was added to over a period of centuries, and seemingly interred alongside the people were the remains of at least eight sea eagles, perhaps as many as twenty-three. The site is complicated by the fact that the eagles date *not* to the Neolithic, but to the Bronze Age – scientific analysis has shown them to have died as much as a thousand years after the humans.[8] But they're not isolated finds: sea eagles have been found in other tombs, and some were certainly buried at the same time as the people. What conclusions might we draw from the burial of wild birds with people?

The Tomb of the Eagles was once an accessible tourist attraction, its evocative name and location drawing people to explore inside, but like many small businesses operating in a similarly limited space it was an unfortunate victim of the 2020 pandemic. I was lucky enough to visit with my family on an overcast day one autumn, when we took the opportunity to explore the islands a little around a conference I was attending. We spent time looking at the exhibition in a small building near the farm, then walked through the flat and treeless terrain to the tomb itself, on the eastern edge of the craggy, irregular island. Accessing the cairn was a feat in itself, through a small opening and along a three-metre tunnel; we lay on a small trolley, hauling ourselves in with a rope hanging from the ceiling, one at a time. Once inside, modern skylights illuminated a series of small chambers with similarly low entrances where the bodies had once been placed. The children were quite taken with the tomb's cave-like quality: fairly cramped, damp, but not really a place where anyone living would be inclined to spend much time. Although these kinds of burial sites seem to have been left open for a time before they were closed off, they were not really intended as places for people to gather in. The most memorable thing about the tomb is not its form or what was in it, but what lies just outside it – the tufty edge of the eroded cliff, the rocks and the seemingly limitless expanse of sea stretching eastwards to what we now know as Norway. This, like so many monuments, was as much about *where* as *what*.

Its modern location is little changed from its prehistoric one. There may have been some coastal erosion, but this stack of stones and bones would always have commanded a position on the edge of the world, in the nesting and feeding ground of one of Britain's biggest raptors: the sea eagle. This bird has had a complicated history alongside us. In the seventeenth and eighteenth centuries it was feared for its threat to infant humans and animals alike – unlike scavenging corvids, the eagle is more than capable of carrying off a toddler. With an extraordinary wingspan of over 2.2m, the female can weigh up to 7kg and carry at least her own bodyweight in flight. It is easy to understand why the presence of eagle bones in the tomb indicates that the animals were totemistic, representing something that was closely tied to death. Some believe that the bones are incidental and the birds were using the tomb as a nesting place, and it's true that there is an inherent danger here that our *current* view of the wild is what leads us to believe these birds had a special place in later prehistory. After all, when we find burials of dogs we think of them as companions, but the remains of cattle we think of as food. But the fact remains that scavenging and predatory birds would have had an active role in helping to break down bodies, and their persistent presence in myth and depictions on objects over time do lend themselves to an interpretation that they were something more symbolic. Birds were not property like sheep or cattle, and were not companions like dogs. They had their own lives and territories independent of humans, but overlapped with us in life and in death. They lived *more than human* lives, unlike those animals we kept within our sphere of control. Are the burials of these beings evidence that the wild, even in later prehistory, was already something that was seen as better than us, without us, or is it the wishful thinking of a modern mind shaped by fairy tales?

Between the wild and tame

Ready answers to questions that are so separated from us by time are difficult to find, and there are other beings that muddy the waters here: those that had loose equivalents in the wild, although they were sometimes more like reflections in a curved mirror. Pigs were, like cattle, domesticated in the Near East and appeared as one of a suite of things that heralded the beginning and spread of the Neolithic. By the time these Near Eastern pigs made it to modern France, European wild boar were also being domesticated, and once that happened they outcompeted the Near Eastern domesticates and entirely replaced them. They are fairly common on sites from the Neolithic and Bronze Age but much less so in the Iron Age, except where they are found in high concentrations on particular sites. This seems to be because they were easily raised for feasting, which as we've seen was important to prehistoric culture but particularly so in the Iron Age. Pigs are, however, greedy and destructive, and would have been in competition for cultivated land, especially in areas of dwindling woodland where they could forage the autumn tree fall and among the roots. In short, rearing pigs was an investment, but it was also probably not just because of their easy meat that they were considered special.

By the time we began to use iron many animals were well into their journey of domestication, and we kept managed populations of sheep, cattle and horses. The pig was one of only two animals that continued to co-exist alongside their much wilder, bigger and more dangerous counterparts. Where the dog had the wolf, the pig had the boar. These were also two of the only animals that could really pose humans much of a real threat in Britain – almost all of our megafauna of old was gone, though it's possible bears clung on in very small numbers. Porcines represented either side of the wild/domestic divide, and although iconographic depictions are fairly rare, those that do exist focus on the boar over the pig, as a warmonger, a danger and an aggressive force. They sometimes topped the carnyx, a type of war trumpet that was held up into the air and called to excite troops or incite the beginnings of violent battle, and they are also found on weapons and coins. Despite their ferocity, the young of wild boar are reasonably easy to tame, and there may have been some short-term domestication activity through time, even after populations of domestic pigs were established. The boar survived, unlike the aurochs, perhaps because there was just enough habitat for them and they do not seem to have been hunted very much, if at all. What few archaeological remains there are also suggest they even grew in size, possibly benefitting from the new regimes of land management that provided concentrated forage.

Wild introductions

Management was not confined to domesticated stock, as people were clearly managing certain *wild* populations as well, including some that were deliberately introduced as wild wanderers. Both wild boar and red deer were imported to Ireland in the Neolithic. There is no evidence of them existing on the island before this point, and semi-managed populations must have accompanied humans from elsewhere. In the Scottish islands, too, it seems that red deer were taken across the sea and formed populations that survive today. This was despite the fact that communities were settled and routinely relied on domesticated animals for their food. Deer remained important, clearly so much so that it was worth shepherding them across the sea to form a semi-managed herd. Antler was used in tool-making far beyond the Mesolithic, and their hides may have been used in clothing and shelter. What is really striking about these red deer is that those from the Inner and the Outer Hebrides appear not to be related. Deer can swim up to 7km, and this explains why the genetic signatures of deer on the Inner Hebrides and mainland Scotland match. But those on the *Outer* Hebrides, much further away and almost certainly introduced by people, appear to be related to a currently unknown population that originated from much further afield.[9] This would make them the remnants of something a little obscure, a kind of half-wild species that accompanied people on a long journey but was never domesticated like other animals. It may seem surprising given that red deer remain a core identifying characteristic of 'the wild' in Britain, but they're not alone. Another icon has a surprising prehistoric origin story.

As late winter yields to warmer weather, some people are lucky enough to witness one of the springiest rituals of all, backlit by the sun through the mist: the boxing brown hare. Standing tall on their hindlegs, the female hare bats away the over-enthusiastic amorous advances of the male. Most often hares are spotted in the open fields, sitting low to the ground over shallow depressions in which they birth and protect their young, though they can be glimpsed darting around trees in the woods when evading predators. They are one of our most well-loved mammals, depicted as running, leaping or moon-gazing, and woven through folksong and folklore into the history of our land. They are particularly prevalent in tales of witches, who are often able to take the form of the hare to escape detection.

The hare is stalwart in our landscape, like the hedgehog and the water vole, held aloft as a native worthy of the highest protection. Indeed, when the clothing company Joules disputed another brand's attempts to use the image of the hare in their own logo, the paperwork stated that the sitting hare was adopted because it is 'emblematic of the countryside'.[10] But recent studies show that the brown hare first made its appearance here in later prehistory – it is not a mammal that is native to the British Isles at all. It outcompeted the incumbent mountain hare, which was once more widespread but is now found only in the highest parts of Scotland. Current analysis points to an introduction in the Iron Age, with a steep increase in population numbers in the Roman period, before a drop in the early medieval period again. There may well have been captive populations, but it is equally possible they were introduced *as* wild to populate the landscape, like the deer and boar of Ireland. Unlike the pig, the sheep and domesticated cattle, the hare doesn't seem to have been eaten. Caesar suggested that Iron Age Britons considered it taboo to eat the hare, and tales of Boudicca claim she released a hare from her clothing during a battle with the Romans, invoking the war goddess Andraste. Hare skeletons are most often found completely intact, with no evidence of butchery or processing, and in a range of associations with other wild animals such as foxes and deer. Despite this, iconography that features the hare remains extremely rare before the Roman period, and this has led some to suggest that taboos surrounding hares extended to their depiction. Perhaps most surprisingly, the hare doesn't seem to have truly naturalised and become really at home here until later into the medieval period.[11] So, its introduction was not about providing stock for food but must have been bound to something else.

The introduction of all these new animals helped to shape the way we think about the countryside and revolutionised the way we moved around it – on horseback and on wheels, but also as we settled into year-round dwellings that catered to our needs. They necessitated and went hand in hand with another major element that materialised over thousands of years. We wholly reshaped the land, structuring it according to new ways of life, causing effects that have rippled into the now.

From the forest to the field

The moorlands of Britain are often thought of as a wild space. We think of them as untamed and roamed by beasts unencumbered by barbed-wire pens, and they are often sites of dramatic literary happenings, as well as real-life tragedies. Certainly, the moors are challenging environments, and traversing them or even spending the night sensibly requires specialist equipment and knowledge. The weather, with its disorientating mists, can roll in up the valleys. For some this wild framing of such a space renders it off-putting and inaccessible, as not *for* people (or perhaps just not for them). But these windy hills now covered in heather, bracken and fern and all variety of mosses and bryophytes used to be our upland woods, except in their highest and rockiest extremities, and they are more human than you may expect: moorland didn't really exist here until we undertook large-scale forest clearance for farming the animals and plants we'd brought. And as our thirst for crops and a sedentary lifestyle intensified, so did our creation of new habitats.

Instead of the unenclosed and unmanaged land we'd had in the Mesolithic, a new type of landscape was fabricated in its place, in the form of hedgerows and walls, fields and pens – barriers against wild places. Everything became *edgier* – not in the sense that it was more dangerous, but because anthropogenic compartments were created where habitats came to an abrupt end: perhaps the edge of a wood, or where a wall divided wheat from grazing land. Today, as we look around our countryside, edges are really all we can see: the straight edge of a copse, a telephone line, the legal boundary between urban and greenfield made visible by the interface of field and housing estate. The zoning that made these edges was a product of later prehistory, but also had unintended consequences. The intersections became habitats in and of themselves – the hedgerow with apples seeking to expand their territory out of an orchard that has now disappeared, or the occasional glimpse of barley beards paling on the road verge.

These edges developed to serve the organisation that was required by the animals and agriculture that had arrived in the Neolithic, and landscape boundaries necessarily became formalised. It brought a permanence to things that had been little seen before outside of geology. The farming of the Neolithic was reasonably small scale and perhaps short-lived in places, but it was significant. The early cereals introduced here were emmer and spelt wheat, and barley. Plough marks are very occasionally found underneath Neolithic barrows, showing cereal cultivation prior to the burial mound being constructed. It was an important and intensive undertaking, not least because this was farming that used cattle to pull the ard, cattle who were thus required to be not just domesticated but also trained for a job. It is difficult to overstate how different this way of life would have appeared to those populations who survived by hunting and gathering, and it was only amplified over time. During the Bronze Age, farming intensified, driven by a complex interplay of factors such as population pressure, improved agricultural technology and increasingly complex social structures. Agriculture expanded onto higher ground, aided by favourable climatic conditions in the Early–Middle Bronze Age, and the first extensive unenclosed settlements appeared. These consisted of the now-iconic roundhouses and their nearby fields, some with attached plots that we'd think of now as gardens. This new permanence had a knock-on effect on the biodiversity that was able to thrive in our land, and that still thrives today.

The intensity of this Bronze Age farming in places like the moors was not matched in later periods: ultimately, it was always marginal land, a fact exposed towards the end of the period by a cooling climate. Traces of moorland fields and houses have therefore escaped flattening by later intensive ploughing, giving us some of the best visible evidence we have of prehistoric farming. Giving a date to most features of this type has traditionally relied on attempting to reconstruct complicated relationships across large landscapes, looking at how boundaries and places intersect or obscure each other to build up an idea of how they developed over time. The Dartmoor reaves, low stone walls that run in straight lines across kilometres of moor in Devon, have been dated by form and excavation to the Bronze Age. These are not subtle and faint traces of buried archaeology that are visible only during a dry summer; they are still very obvious on the ground today.

Although I used to live nearby, I had never really paid much attention to the reaves and had certainly underestimated how rare it is to be able to see such a network of prehistoric settlements laid out across a large area. So, as I revisited moorland archaeology in my research for this book, we piled into the car for a few days of boundary-hunting one February half-term, when the grasses were low and the frost was friendly. As we drove over the moors in search of some accessible reaves, my youngest daughter told me it looked like a desert, and she had a point. In the low February sun, the grass was sometimes pale yellow, the other vegetation tending to honey and caramel. But she was also bang on in some other respects. Certain moorland habitats are ecological deserts, devoid of the richness of their former biodiversity because of much more recent overgrazing, mining or quarrying. On the other hand, there are some areas that abound in mosses and bryophytes, lichens and shrubs, that were only able to develop because of forest clearance in the first place, and that are largely held in place by the grazing of sheep, cattle and ponies – all introduced by us.

Taking the time to stand and look at reaves and the grid-like pattern they form, I began to appreciate how unusual the Dartmoor reaves are. They pay little heed to the shape of the underlying land, plunging into valleys and across streams, steadfast in their straightness. They can be seen disappearing over the horizon in all directions, occasionally interrupted by the remnants of roundhouse walls. Now devoid of all but a few tourists in the low winter sun, thousands of years ago it was an entire landscape of building, farming and living. Clearance of the old forests continued into the Iron Age even as the Bronze Age arable boom began to dwindle, resulting in the grassy moor we have now.

While burning seems to have played a part in the maintenance of openness, the presence of the plant ribwort plantain in palaeoecological samples reveals that for at least the last 3,000 years it has been grazing that has been the major force in holding the environment in stasis.[12] Many would recognise ribwort plantain even if they don't know its name: rosettes of long, ribbed leaves, with curious flowers that have heads like small pine cones and tiny flowers shooting out to form a kind of scraggly tutu. It is commonly found in urban and suburban areas, growing along the edges of pavements and in other disturbed ground. It may be easy to overlook, not the most rare or beautiful of plants, but the reason it is so common is that it's a species that has been thriving on human disturbance for thousands of years.

Sunlit hedges

We don't always need large areas of landscape like Dartmoor to date the divisions that encroached upon our lost wild spaces. It had long been suspected, for example, that some hedges in the west of Cornwall were prehistoric, but Hooper's Law (using the number of species in a hedge to work out its age) doesn't work there. Cornish hedges are not just a linear collection of plants but are *constructions*, made up of earthen banks with stone facings, topped with trees like elder, blackthorn, oak and ash. The banks alone are typically around 1.5m tall and their unyielding sides regularly catch out drivers who are unused to narrow lanes and the frequent need to reverse long distances in the face of an oncoming tractor. Wonders of biodiversity, their stone crevices form microhabitats, with damp sides suitable for a host of wet-loving plants and ox-eye daisy tumbling from the cracks. In late spring, they run through the hills as ribbons of vertical wildflower meadow, all cool hues of bluebell, scabious and ground ivy, and pinks of rosebay willowherb, foxglove and red campion. The hedge base forms a sheltered corridor for small mammals like shrews and can provide reptiles such as slow worms with a warm shelter. As a vertical habitat, these hedges are also immensely important to solitary bees and songbirds.

The unusual form of the Cornish hedge makes it an ideal candidate for the use of a particular type of scientific dating that is, remarkably, able to tell us when sunbeams last touched the layers hidden within. Optically Stimulated Luminescence (OSL) works on the basis that buried sediments absorb radiation from the sun over time, which is exhaled as a burst of luminescence when exposed again to light. By comparing how much radiation is required to create the same level of luminescence in the lab, and comparing that to the background level of radiation in the environment, we can obtain the approximate date when the sediment last saw the light of day.

Using the technique on the hedges of Bosigran, west Cornwall, has confirmed that they are indeed prehistoric, constructed by Bronze Age farmers around 1690 BC.[13] The hedges here, then, scraping cars, obscuring sea views and easily dismissed as mere boundaries, have lain in place and delineated specific areas of land for over 3,500 years, communicating to generations of people the concept of the field. And they are not the only features that have been dated this way. Lynchets – earthworks appearing like shelves along steep hills, formed of strips cut into or built up from the slope that allowed agriculture in places where it would have otherwise been difficult or perhaps impossible – are also notoriously difficult to date with anything beyond a vague guess and have likewise recently benefitted from the OSL treatment. In Northumberland, recent work has shown some of these lynchets to date to the same period as the Bosigran hedges.[14]

The Bronze Age seems not to have been characterised by the small-scale clusters of huts with a few cows, living in harmony with the surroundings, but was a period of huge demand on the land that magnified the removal of the wildwood that we had lived within and alongside in the Mesolithic. Physical barriers were created that have persisted in time and in meaning, as a way of protecting our domesticated plants and animals from whatever was outside the bounds. More so, they represent the disconnection from a wild way of life to a fully settled and domesticated one that has followed us ever since. The moors are not and really never were wild in the way we want them to be, but are instead the end result of a complicated mass of interactions between people, geology, plants and animals. The state of their biodiversity matters enormously, but we are key to its creation and maintenance. Leaving the moors to the ponies and whatever wants to grow there is not what they need, but rather we have a responsibility to maintain a more carefully considered and curated community of plants and animals that works best for the future.

The road to nowhere

For the moorland reaves at least, the fields they enclosed seem to have been used for a few centuries before the method was abandoned when the climate deteriorated. However, the uplands continued to form an important component of grazing regimes where cattle were driven onto the hills to graze across large areas in summer – and in later periods we would return to use them in more intensive ways. The movement of livestock between upland and lowland helped to maintain the new open character, but also the routes that led people and animals to and from this valuable resource.

Most of the roads and tracks, paths and lanes we encounter today are static, fixed lines that we follow through reliable routes, but it wasn't always so. The old ways were much more fluid and shifted over time as new desire lines were followed, perhaps because an area got too muddy one winter or because a tree had fallen. The tracks led to and from places that were fixed to a degree, certainly, but the routes traversed whatever land was the most convenient at the time and was not constrained, as we are now, by the complexities of private property and the dangers posed by traffic or railways. When the conditions are right, aerial photographs sometimes show the shadows of these old routes, lines coming from all directions and converging on an old pond or a pass through the hills. We'll never know where most of them were going.

It is always tempting to assign very ancient dates to the routes we still use. We are familiar with the Roman roads that have survived in many places, running as straight today as they were when they were made. But there are also routes that are popularly thought of as droves used by 'the ancestors', such as the Ridgeway in Oxfordshire. Like many high roads along the ridges this was long thought to be prehistoric, leading the way across the grassland high above an impenetrable primal forest below. Aside from the fact that we know that any lowland forest was far from uninterrupted anyway, those who have looked more closely at this route also now believe it to be Roman or even later.[15] Some rare examples are truly ancient – occasional tracks in Ireland and the east of England seem to have been local droves between upland winter and wetter summer pastures and have been shown to be most numerous in the Bronze Age, when the movement of livestock over distances was at a peak.

Sunken lanes, or 'holloways', are often given the romantic treatment, imagined to have been worn as a scar into what was once an untamed and wild land. Holloways are actually very difficult to date, as many have formed because of natural variations in the underlying topography before coming to be adopted as a convenient travel way by people. On the chalklands they make use of naturally occurring combes, steep valleys that have often been formed by 'spring sapping', where water emerges from underground channels and erodes the bedrock away. Elsewhere, the holloways cut up through natural folds that resemble fondant draped over the edges of a cake, sitting above what was the river valley. These elegant landforms are known as flutes, made by the freeze–thaw process long ago. On stormy days the rain runs off the hills and torrents down the road, and when it does it's very easy to see how the steep sides that tower above the road were formed and how any movement by people and animals was exaggerated by the force of the water. There is no doubt that many routes have been in use for a very long time, but who the first groups were to continually use them is too often invisible.

Sometimes it's the tracks we no longer use that are the easiest to date, but they're often found only by chance. Footpaths are funny and temperamental things, so many traceable only by the stunted growth or complete absence of vegetation. Tracks can come into being in a matter of hours, given enough footfall, and disappear after a few days of plant growth. If it were not for the constant shaping of vegetation, the vast majority of our footpaths would have disappeared years ago. So many ways have been lost that will never be recovered. On occasion they peek through after excavations, but often these don't show grand narratives of migration and long-term use but much more personal vignettes of local movement, of people going about their day in the past. The oldest constructed path uncovered in Britain is the Sweet Track in the Somerset Levels, a single-file trackway raised across a marshy area and in use for perhaps only a decade in around 3800 BC. Running for 2km, it facilitated the movement of people between areas of dry land. Sometimes, though, tracks may have existed only to allow access to wet places in the bogland, where special goods such as the flesh-hook described above could be deposited. By the Iron Age, these short-lived tracks tail off, in line with declining evidence of wetland exploitation and a singular focus on settled agriculture. Still, though, the wet places seem to have held our fascination, and possibly fear, for thousands of years.

Between land and sea

In the late 1990s an amateur archaeologist, John Lorimer, was crabbing on a sandy north Norfolk beach, peering into the crevices among the interface of ages-old peat and silty sand that had been exposed by the retreating tide. In recent weeks he had found a bronze axe head on the same stretch, assumed to have come out of the prehistoric peat with tidal erosion. On this particular day, he noticed several timbers visible on the beach, including what appeared to be the upturned stump of a dead tree that had washed in. But upon closer inspection, the stump had not ridden the waves into place but was revealed to have been deliberately hoisted into position in the deep past. Over the next few months, as the water took away more of the surrounding sediments, the tree was found to form part of a larger monument, embedded in the centre of an egg-shaped arrangement of timber posts. During the summer of 1999 rapid excavation took place on a now legendary dig that revealed the first survival of its kind in Britain: an intact prehistoric timber circle, a monument that was to become known as Seahenge.

Were it not for Lorimer's abiding interest in archaeology, the posts and the tree may well have disappeared that same year, gnawed out of the sediments by each tide and washed out to sea. Archaeologists had to work quickly and in short bursts to excavate and record the timbers and sediments before the sea could reclaim the visible layers and destroy the relationships between the different archaeological components. It was a controversial decision, but the monument was removed in its entirety to enable specialist analyses on the surviving wood and, for what it's worth, most would agree it was the right course of action. It is very difficult to find parallels for the story the wood had hidden in its grain and its tool marks. In later years it was discovered that Seahenge had a sister monument very close by, a second timber circle. This one has been recorded and left in situ for the tide to eventually unmake.

We now know that Seahenge and her sister were constructed from timbers felled in the summer of 2049 BC, in the Early Bronze Age. The original monument was of an upturned oak stump that had been stripped of its bark and dragged into place using a rope of twisted honeysuckle branches passed through a hole drilled into the trunk. This central tree with its roots reaching into the sky was surrounded by over fifty tightly spaced split trunks of oak, positioned with their bark facing outwards. The interior of the circle would have been almost inaccessible and shielded from view, but for the use of a single naturally forked trunk that allowed entry for those that could squeeze through the gap. What these timber circles were used for is much debated, interpretation hampered in part because, ironically, much of the evidence we might recover in a dry and static environment – the layers and their composition – does not survive the challenging and constantly shifting tidal interface. The timber has been preserved by the water, but the dynamism of the sea washes away much of the other evidence. Timber circles more generally across Britain seem to be associated with the dead, and it may well be that the upturned oak was for holding a body for excarnation, bringing it into contact with the wild animals we have met above that defleshed the corpse before burial, cremation or other method of disposal.

When the timber circles were constructed, this stretch wasn't a wet beach and the monument was not a hasty arrangement between tides. Plant remains recovered from the dig show that the area was actually an inland saltmarsh of samphire and sea milkwort that only much later was claimed by the rising sea.[16] The oak used in the monument must have come from elsewhere as this was not a woodland. Yet the appearance of Seahenge, with its tightly spaced timbers, would have rendered it like the trunk of a giant tree – one that was out of place. This was a period of climatic fluctuation, with harsh winters that negatively affected the growth of oaks that were already suffering from the pursuit of agricultural land. The monument would have resembled the trees that were slowly disappearing from the forests, and for this reason it has been suggested that timber circles may have been intended to represent woodland itself,[17] used as places of commemoration or mourning not just of humans, but of environments that were being lost.[18] It is possible that Seahenge was a prehistoric locus of ecological grief. No-one lived here; there is some evidence, particularly later, that wetlands were actually avoided in many circumstances, and this environment was perhaps the wildest there was left, in a land that elsewhere had now seen two thousand years of domesticated animals and farm crops. A monument to grief, uniting the woods and the wet and a diminishing wild with that which remained.

Supernatural wilds

Wet places as an enduring wild in prehistory, particularly the bogs, mires, fenlands and marshes, portends their place in later stories and folklore, reflected in the unease they often inspire even now. Many people in prehistory seem to have been content to live near them, but not necessarily venture into them or heavily exploit their resources, except for particular activities or in rare cases when settlements were constructed on lakes. In Ireland, there was some short-term occupation at wetland sites in the Neolithic and earlier Bronze Age,[19] and elsewhere, for example, there are places where it looks as though humans and beavers alternated, using the habitat in a kind of push and pull that represents time-limited human impact. At the Bronze Age site of Must Farm in Cambridgeshire, a catastrophic fire destroyed a wetland settlement in a single event, preserving textiles of both cultivated flax and wild nettle, with beavers apparently attempting to use the timber that was left once the people had deserted. Even in places where people stayed across the longer term, at some sites in the Fens as well as the famous lake village at Glastonbury, Somerset, wild animals are often recovered in very small numbers – small enough that perhaps access to these places was restricted to specialists and/or outright feared. The people who could engage with the wild rather than just crossing its space were special or unusual. Rather than taking much from the bogs and the marshes, they deposited things in them, giving back to the wild spaces that may have been perceived to carry some kind of power.

The evidence of this practice occasionally rises to the surface, quite literally, and can be famously quite gruesome. In 1984, peat cutters working near Manchester were digging and hauling sods up onto their conveyor belt when they made an unexpected and rather grisly discovery: a disembodied head. So well preserved were the skin, flesh and hair that it was assumed to be a recent death and even led to one man confessing to his missing wife's murder some years before. Except this body was not twenty years old; it was closer to 2,000. It was not to be the last set of remains found in that bog, and only a year later a more intact body was found and dubbed 'Lindow Man'. These are now known as 'bog bodies', with the Lindow finds dating to the very end of the Iron Age. They are not isolated finds: we have many examples from across Britain and Ireland, as well as across northern Europe. They are fascinating for the things that the anaerobic environment can preserve. We know, for example, that Lindow Man's last meal was of a charred flatbread. Though four grains of mistletoe pollen were also found and have been oft cited as evidence for a ritualistic last meal on the basis that Pliny, the Roman writer, stated that 'the Druids' consumed it ritualistically, they may just have easily been blown onto his food.[20]

Bodies aside, bogs, in particular, are special and ecologically valuable places. These expanses of verdant green, russet and yellow are peppered with inky pools that seem endlessly deep; the ground may lull you into a false sense of security before gentle ripples and quakes in the surface give way as you plunge into an abyss. Bogs can shrink and grow; they are affected by rainfall, and if stripped of too much peat are capable of bursting. They are formed from decaying sphagnum moss, the name for a collection of plants that form a springy carpet over wet and acidic soil and make the ground appear as a forest in miniature. This decay turns bogs into carbon traps and makes them suitable environments for the preservation of all sorts of things. Bog bodies are the most famous, but are certainly not unique. The acidic environment preserves wood, flesh and food, all things that rarely survive in the wider archaeological record. Among the stranger finds are what is known as 'bog butter', a waxy substance that has been found to date to as early as the second millennium BC and as late as the nineteenth century AD. Wrapped in bark and thrown into the bog, it seems to have begun as a method of preserving the excess of successful dairying regimes but, given the deliberate deposition of both weapons and gold in bogs, is thought to have been imbued with something else too. Later, it was a way of storing and infusing extra flavour into preserved dairy, and was eaten as something of a delicacy.

It seems little wonder that bogs have developed a reputation as unfriendly and dangerous places full of spirits and symbolism, home to boggarts, fairies and lost travellers. Given their propensity to preserve things for thousands of years, it is highly likely that prehistoric people also came across ghoulish sights preserved in the peat, and would have certainly been familiar with its ability to swallow up whole animals and people who strayed. It is entirely possible for someone or something to disappear in the blink of an eye, any ripples in the surface of the pool obscured or quickly swallowed by vegetation. It is the sphagnum moss that causes and enables this deception; a friend who grew up surrounded by raised bogs described to me once how it knits together over both water *and* solid ground, making them appear one and the same. Her childhood memories of the bog are dominated by a fear and fascination of an entity that could so easily snuff out a life in total silence.

Will o' the wisp

Another legend associated with the mires are the faint and beguiling lights that danced through the air, mimicking a torch: will o' the wisps. These spectral glows are generally accepted to be real rather than mythical phenomena. Yet, we have no entirely satisfactory explanation for their existence. Science suggests that they are a product of the oxidation of various gases that occurs as the result of decay, that there is a spontaneous ignition of methane over the surface of the wet ground.[21] But there are issues – the temperatures needed to cause the reaction are too high to occur naturally, and although attempts to recreate wisps in lab conditions *have* resulted in something resembling a wisp, these lacked their famously blue glow.[22]

A further challenge is that wisps seem to have become extinct: there do not seem to have been many, if any, reliable sightings in the last century. We cannot study them because they have gone, and for this there are two intriguing explanations. One is that the intensification of use of these landscapes has meant so much draining and disruption that their natural processes no longer function in such a way that produces the conditions under which wisps occur. In this scenario there are no longer places wild enough for these luminescent beings to come to life, as we've tamed even the supernatural. Another explanation is that wisps were *always* a human phenomenon, born not of our minds but of our flames – that wisps were made by the traveller crossing the bog, their torch igniting the marsh gases, which took on a life of their own and danced away. They have disappeared because we are no longer preceded in our wanderings by a flame that lights our way and brings them into being. In this sense, they were the very product of human foray into the wild: something created by humans, but which exacerbated the divide between us and the natural environment.

Between the sky and the sea

I want to end this chapter by thinking a little about the interplay of animals, humans and the environment in one of the sites above, Seahenge, in the Early Bronze Age. By then, the wild had a part to play in some, but not all, of daily life. Imagine for a moment a figure breaking away from a seasonal settlement, following a path through the oak wood and out towards the marshes. As the oak gives way to alder and willow, the paths turn to brushwood, thrown down over moist ground and held in place with stakes. After a while she comes to a stop where, through the reeds and the sedges, the dark, damp bark of split oak trunks screen the view to what lies beyond – an upended stump with its roots turned to the evening sky. In recent months the body of one of her community was carefully brought onto the marsh, manoeuvred through the fork-shaped entrance in the circle of oak planks. It was placed, naked, onto the natural platform formed by the twisting roots, given to the wild beasts to strip it of flesh before the elements worked to bleach the bones. Every so often a raven flits briefly into view, hopping over the roots as it explores the body for any remaining meat, indicating that the time for removing the bones for the next phase of the mortuary journey isn't yet here. The bird's caw rings through the air, and with the sun bedding down below the horizon the surroundings cool and a mist forms over the surface of the marsh. The figure stoops to retrieve a torch that will see her safely home. As she turns away, the spirits of the dead waft from her torch as wisps of blue to dance around the old oak in the haze of dusk.

Years later the site was slowly claimed by the sea, eventually disappearing. But so, too, were many of the woodlands, claimed not by the sea, but increasingly struck down to make way for the fields and newly enclosed settlements that appeared throughout the Bronze Age and were cemented as a way of life by the Iron Age. We do know that the wild played its part in the cosmology of these people, in their iconography, in their beliefs about the supernatural world and the impact it could have on the visceral experience of day-to-day life. But attention had turned from surviving on what we could gather to what we were able to make. We had become producers and owners, taking control to begin shaping plants and animals into what we needed. Around 4,000 years after the gathering at Coneybury we had formalised our way of life into that of sedentary villages and metalworking, more likely to be fighting other people for survival than dangerous beasts beyond the boundaries of domesticity. It was into this largely tamed land that a power across the water, the Roman Republic, turned its attention two thousand years ago.

3

EXOTICISM

(ROMAN, 43–410)

Bad wolf

The Romans have a well-deserved reputation for opulence in food, art and buildings. When visiting a Roman villa site or museum, you might be lucky enough to see exquisite mosaics featuring intricate patterns, nuanced colour and shading, exotic fruit and animals. But like famous cases of bad taxidermy and art restoration, the Romans also made mistakes and bad mosaics happened. There is a wonderful example in Leeds City Museum, a mosaic found in Aldborough in North Yorkshire. It's crude in design, featuring a wonky border and a tree that has been squashed around the inner edge; however, it's the animal in the centre that is the real horror. The creature appears to have the back end of a horse, complete with high rump and characteristic tail, and a head turned towards the viewer featuring a large grin with many pointed teeth, a strikingly enormous nose, round eyes and pointed ears. Two small figures underneath its belly suggest this strange creature is *supposed* to be a wolf, the figures representing the twins Romulus and Remus from the legend of the founding of Rome. Comparing this mosaic to others across the Roman Empire, it is difficult not to conclude that the artist simply didn't have any idea of what a wolf looked like. Although there are stories of wolves surviving in Britain until the eighteenth century, we have very few verified wolf remains in the archaeological record. In part, this may be because they are skeletally so like the domesticated dog; however, it is also entirely possible that before or during the Roman occupation of Britain we had already hunted the wolf to extinction or else caused numbers to dwindle so significantly that wolves were not something encountered with any great frequency. Perhaps it was the power they held as the wildest disruptors of domesticity, the devourers of livestock, that meant wolves were imagined to be living on in the countryside for a further 2,000 years.

Putting aside whether or not there were wolves roaming the rural landscape of Britain when the Romans are said to have arrived in AD 43, what they landed in was not a thing of untouched wilderness, but a landscape that was already a mosaic in and of itself: scattered settlements and farmsteads, surviving but largely managed woodland and the kinds of moorland that we'd recognise today, created and then abandoned by agricultural endeavour over hundreds of years. In the first century AD, a large part of the British Isles became the farthest north-west tip of the Roman Empire, but just as it wasn't the unending expanse of untouched woodland it's sometimes imagined to be, neither was it culturally isolated from Rome in the first place – in some ways, the arrival of the Neolithic 4,000 years before was just as sudden and dramatic. Iron Age Britons were already part of a trading network that meant that even in the first century BC there was significant contact between the south of England and Gaul, the region now covered by France, Luxembourg and Belgium. Still, the Roman occupation of Britain revolutionised many parts of the landscape and the way people lived. Among the introductions were road networks and bathhouses, olive oil and military camps. There was also a sea change in the types of food being consumed and the forms of entertainment that people enjoyed. The Roman approach to wild things, particularly animals but also land, was very different from that of prehistory, and had its own great and lasting impact on the wild of Britain.

We'll start this chapter by looking at the kinds of evidence we have, and at some of the ways in which the idea of Roman Britain as a coherent whole is a myth. Then we'll turn to the changes and introductions that have echoed down the years, at the plants, animals and *attitudes* that have helped to form our lived environment today. There are a great many apocryphal stories about what the Romans did for us, and what they left here that has survived. With archaeological and genetic techniques, we have been able to dispel some of these as myths, while with others the results have lent credence to the oft-repeated assertions.

The evidence

Because of the enduring popularity of the Romans, coupled with the fact they established cities and more infrastructure than had been seen before, there is a huge wealth of archaeological material from the period. But it's also the first for which we begin to see written evidence. There are surviving works of literature that describe to us events, ways of life and the topography of Britain before and during Roman rule. In these we find the first naming of Britain, referred to as *Prettanike* in early Greek sources, and then *Bretannia* or *Brittannia*. All of the isles were sometimes included under this umbrella, but Ireland was also *Hiberni* or *Iris*, and Scotland *Alba* or *Albion*. While very useful, the written sources cannot be relied upon in isolation. Many, such as Caesar's autobiography and Tacitus' *Agricola*, the latter written around AD 98, were about presenting a glorified version of a specific person rather than a balanced account of another land. There was a certain recycling of information that may not have been entirely accurate in the first place, and first-century BC Greek writer Diodorus Siculus partly based his account of Britain on Caesar, as well as other works that are now lost. In Tacitus' work, the inhabitants of Britain were presented in a similar vein to most of Rome's neighbours: uncivilised and with poor diets, aggressive and violent 'brutes' who clothed themselves in animal skin or simply went naked except for their body paint. Other accounts were similarly judgemental. Herodian wrote of an atmosphere of gloom caused by the dampness of the land, and Dio describes inhabitants of the north living a somewhat nomadic existence in waterless mountains.[1] Greek writer Strabo described how the British were *so* barbaric they drank milk but were incapable of making cheese, and were unable to garden or even farm effectively. Britain is said to have had exports, especially large dogs that were suitable for both the chase and in war. Writers were also quite taken with the relationship between land and sea, with the tides causing river swells unlike anywhere else they'd encountered, and with the wetlands and marshes that accompanied this. It was an important component of the myth-making of Britain as a wild place, at the very edge of the known world, bordering unchartered seas, and perhaps the literary genesis of wildspeak itself.

These texts have survived over time by having been copied and copied again, so what has come to us is mediated by the hand of others. However, there are archaeological examples of Roman writing found in Britain. They don't take the form of big sweeping histories but are small records of individual events, tax records, curse tablets and even, as we'll see below, land transactions. This writing was being produced early in the Roman occupation; the earliest that archaeologists have uncovered so far were found in London. These are known as the Bloomberg Tablets as they were found during excavations ahead of the construction of new company headquarters between 2010 and 2014. More than 400 wooden tablets had survived in waterlogged conditions. They were covered in wax and designed to be written on again and again by smoothing away the wax for reuse. Within them, both in their text and their materiality, are hints of the wild of Roman life. Almost all of the tablets were fashioned from imported coniferous wood, mostly silver fir, and seem to have been recycled from barrels that were probably used to import wine or oil. Although the wax has largely disappeared, the stylus used to write on the tablets left impressions in the wood underneath, and it was from this that archaeologists were able to extract some of the text. This showed small details, like the trade of beer or accusations of unpaid debt, but also revealed how London recovered from sacking during the revolt led by Boudicca in AD 60–61.[2] The Vindolanda Tablets, similarly, show insights into daily life. These were found at a Roman fort in Northumberland and date to the beginning of the second century AD. Within them are invitations and greetings, and lists of food including gruel, pork crackling and trotters. There are also requests for hunting nets, showing that the Roman army hunted the wild even though they were provided for at the fort.

The edge of empire

Before we can start to understand all of this evidence, written and archaeological, it's important to acknowledge that it was not the same in every part of Britain and that the stereotypical idea of groups of fully Romanised people and soldiers living in towns and forts and *looking* wholly Roman represented a tiny proportion of the population of Britain at any given time. Almost everyone lived rurally, and in different places the character of settlements took different forms. Stories about how 'the Romans', led by Emperor Claudius, arrived suddenly in AD 43 and spread into the landscape tend to make claims about how they didn't make it into Cornwall, Scotland or Ireland. We know that this just isn't true – there were earlier incursions over a hundred years before, when Caesar twice landed. There was a Roman presence in Cornwall, where there are several known forts. In many cases it seems the local populations across Britain were brought into the empire, but in others we know that there was more of a back and forth of troops, ideas and material culture.

Given how close it was to Gaul, which came under the control of Rome in the first century BC after centuries of clashes, it's not surprising that South East England has the earliest Roman artefacts and was the first region to adopt, partly at least, Roman ways. It is particularly visible in the evidence for changes in the types of food and drink people enjoyed, as well as the plates, bowls and cups they used, which start to show up from the first century BC. These were inevitably made by throwing on a pottery wheel, technology which only seems to have arrived in Britain for homegrown production 200 years later. Ceramic *tableware* was really quite new, as previously most ceramics were for cooking and storage rather than forming part of the performance of eating. The arrival of plates, bowls and cups, particularly those that are decorative but mass-produced, hints at new ways of consuming things. It has been likened to the introduction of tea in the eighteenth century, which saw different vessels for different teas and all the ritual associated with tea-drinking. Much of the new Roman material didn't make its way north for some time, but that might not be because transmission was slow but rather because people just weren't that bothered about being Roman yet. In Gloucestershire, for example, people don't seem to have begun to change their clothes or the items they were using until the second century AD.[3]

In the north, boundaries such as Hadrian's Wall and the Antonine Wall were constructed to, at least conceptually, provide a point beyond which Roman control ceased, and they weren't constructed until the second century AD. Both walls, though Hadrian's famous wall in particular, inspire a certain kind of idea that while England had been civilised by the Claudian invasion, Scotland was still living its best and wildest life north of the wall, in the Caledonian Forest, a dense woodland of native pines in which tribes of Celtic warriors roamed. The strength of this image is enough to have been immortalised in works like George R. R. Martin's *Game of Thrones*, where what's 'north of the wall', the Haunted Forest, is something to be feared and kept at bay, requiring specialist knowledge of such ecologies in order to navigate and survive. *Both* of these woods are works of fiction. The Caledon myth was peddled by Roman and classical writers who had not visited but told tales of a land that couldn't be tamed. In reality, there were surviving pockets of pines, but even these had probably been already pushed to the north-west by human activity.[4] The Iron Age of Scotland lasted beyond that further south but with intermittent Roman incursions well into the mid-first millennium; the very fact that the turf-built Antonine Wall was constructed some 150km north of, and decades later than, Hadrian's Wall shows that the boundaries were porous. Borders don't mean that all traces of a culture or intention disappear;

they just define an ideological 'here' and 'there'. While Pliny wrote that in the mid-first century AD legions had not penetrated into and beyond the Caledonian Forest, he didn't really make it clear where exactly that was. At any rate, it wasn't long before Governor Agricola, fresh from leading incursions into Wales and northern England, set his sights on and successfully established forts across much of the lowland area of Scotland. While writers described watery places filled with people who didn't know how to grow crops, even at Hadrian's Wall there is pollen evidence for woodland clearance before *c.*AD 85, interpreted as a result of previous expansion of agriculture rather than performed at the behest of an invading force.[5]

It is true that no part of Ireland was ever taken formally into the Roman Empire, though that doesn't mean that there was no reciprocal influence between both cultures. For a long time, and until recently, much of the material in Ireland that could be characterised as Roman was thought of as incidental, no more than individual examples of items raided or traded from elsewhere. More recently, though, archaeological work has highlighted the similarities between some sites in Ireland and military forts elsewhere, particularly on the east coast, as well as clusters of Roman-type finds around valuable sources of natural minerals, a major income generator that helped to prop up imperial rule.[6]

All this is to say that it is just not appropriate to paint the coming of the Romans as a blanket change across the whole of the British Isles, as if a new veneer of culture were snapped into place. What is described in the following pages is the result of a complex set of interactions between people, ideas, architecture and material that came to Britain from all over the Roman Empire and incumbent populations who adapted to – without necessarily adopting – these new things. There was no single way of being Roman; it wasn't a wholesale transplant of a different way of life onto an existing landscape and population, but nonetheless there were big changes. There were new people, foods, buildings and methods of transport. New species were introduced, but it was not always intentional and sometimes didn't last. There are those that survived only as long as the empire's official control did, while others made their way into daily life for centuries to come.

Rural Roman life

I've worked on a couple of Romano-British sites over the years (technically three, if you count the one under a supermarket in Devon that yielded absolutely zero archaeology from any of the fifty-plus trenches we dug, despite there being a villa recorded in the next field – a memorable dig, for all the wrong reasons). The first was as a fledgling archaeologist during my undergraduate degree, on a Romano-British farmstead in the huge fields and rolling hills of Hampshire in the south of England. It was nothing like how I expected a Roman excavation to be – there were no coins, for a start. In fact, I didn't find a single coin on an excavation for another eighteen years. There were no amphorae or oil lamps, there was no fancy jewellery and certainly no mosaics. I'm not even sure we actually excavated any buildings. What I remember is a boundary or dividing ditch, striking out from one edge of the trench, before turning and running to back into another section. It was a brown line in a stark white base of solid chalk, making it very easy to see. I've not dug on chalk since, and nothing in the intervening years has been as easy to see as that ditch, other than solid walls. The bit that was assigned to me for investigation was a large brown splodge, actually a pit, possibly originally used for something like storing grain. My task was to create a half-section – removing half of the fill of the pit so that we could see how it was built up over time. Much archaeology works on this basis, that all the layers representing different activity will vary in colour and consistency, a little like those jars filled with sand art you get on beachfront walkways, perhaps not as colourful but visible once you get your eye in. And so, essentially, I kept digging the brown stuff until I hit the compacted natural chalk, and what I was left with was the original shape of the pit. In the base, there was an unremarkable small collection of disarticulated animal bone.

This might sound boring but is fairly typical of a rural Roman site that isn't a villa. Most people *didn't* live in villas in Roman Britain, just like most people don't live in manor houses today. A villa was a big investment, normally built of stone and with an array of rooms not found in other houses, perhaps with a hypocaust for baths. It may have had formal gardens and different areas for administration, living and entertaining. Eighteen years after my first Roman dig, I found myself digging an actual villa site, and the experience was completely different. This one was in rural Northamptonshire, on part of the Althorp estate belonging to historian and earl Charles Spencer. The site's Roman life had begun as an enormous stone-based roundhouse, 16m across, indicating that there wasn't always a drastic change in settlement aesthetic from the Iron Age, and sometimes stone wasn't employed at all: its builders just carried on with the wattle and earth roundhouses of later prehistory. The family who lived there seem to have prospered, as their home morphed into a small villa. Some burning, enough to have melted window glass, indicates that part of the villa may have been destroyed by fire, but it was built back and a small tower was added. A few pieces of what may have belonged to a crude mosaic appeared, and it is possible that a full mosaic existed but was destroyed by the ploughing in the field over the years. It wasn't a particularly large or opulent villa as far as these things go, but it was a much more 'Roman-seeming' site than the Hampshire farmstead – there were frequent coins, including tiny copper ones that seem to have been clipped from larger examples. During sieving of a particularly rich ditch sediment, someone even found a minuscule, delicate gold earring; and on the last few days of the dig, half of a beautifully decorated copper bracelet was found, as was part of another made from polished black shale. My favourite find, however, was not a coin or item of jewellery, but an example of decorated bone inlay. Here, animal bone had been polished and etched with ring and dot decoration, a common type of design found in Roman as well as early medieval and Viking contexts. The pieces had been cut into geometric shapes to form the inlay to what we believe may have been a chest of some kind. Some had even been stained the characteristic blue-green that appears when bone has been in long-term contact with copper-base metal. We can only speculate as to its overall design, but so much of archaeology is like this – tiny snapshots that contribute to a bigger story. Among the more usual finds were animal bones including chicken, much pottery, both stamped and painted, and many hobnails from now-lost Roman boots.

The Althorp roundhouse probably came into being in the third century, and was, for one reason or another, deliberately destroyed in the late fourth or early fifth century. There is no continuity here into the medieval period, although this was something that happened with many other places, particularly those that are now well-known cities, such as Cambridge or Durham. Some of the best-known archaeology has come from places that expanded with their Romanisation, like Verulamium (St Albans), Aquae Sulis (Bath) and Eboracum (York). It is notable that two of these place names derive from the natural world: Aquae ('the waters of') Sulis was named for a local deity, Sulis, who seems to have been worshipped at a pre-Roman shrine at these, the only hot springs in Britain. Although she was equated with the Roman Minerva, a goddess of wisdom, poetry and medicine, the name Sulis was retained. Eboracum, too, has natural roots. On one side of a long-standing debate it is cognate with the Old Irish *ibar*, meaning 'yew tree', but in other, similar languages it represents a range of different plants, including rowan and cow parsley. Others believe it means 'boar' and was named for the people who fought like one. All hint at the kinds of things that were important.

Nevertheless, while some towns and cities became major centres, the actual number of people living in them is estimated to have been around 200,000–250,000, less than 10 per cent of the total population of Britain, which is normally estimated to have been up to 3 million at its peak. This might not sound very high when the contemporary population is more like 65 million, but it was denser than at any point prior. Moreover, to put this into perspective, the current rural population when we take away suburban fringes and rural towns is around 5.5 million, less than double what it was at its peak in the mid-nineteenth century. It feels like the countryside was therefore reasonably densely inhabited, made up of farmsteads, villas, roadside settlements and shrines serving travellers.

Tipping the scales

The growing population here required an uptick in resources of all kinds, and this meant a renewed effort to expand settlement and farming again into uncultivated land, or sometimes abandoned farmland, particularly in the lowlands. In order to do this, the whole landscape had to be rearranged, making changes to the way we moved through it and how fields and settlements were organised. A formal road network grew, the most famous of Roman legacies, which became reliable routes between distant places that were known to people far beyond England and Wales. After all, a growing empire required both good communication networks and dependable knowledge of safe ways between important areas. There was also intensifying use of wetlands, particularly in places like the Severn Estuary and the Fens. The Fenlands were transformed through the construction of banks and ditches, which served to convert them from soggy wilderness to something deemed more useful, allowing for peat cutting and extensive salt production. They remained a wetland habitat, but were much altered. Further north there is evidence of Roman canals in the Foss Dyke in Lincolnshire, long thought to have its origins in the first century AD. Canals were useful for water transport; that is, transport both *of* and *on* water. Aqueducts were also created to transport water to military installations at first, and then to major towns, and vast quantities were required for gold-mining operations in Wales. Such constructions were not always great bridge-like structures, but could take the form of smaller leats or even ceramic pipes. What the Roman Empire and its infrastructure did to the wild of Britain was to look upon it as a resource and find the best way to use it.

This led farming away from being reasonably small-scale and mixed-crop towards monoculture, growing a single crop, something that is quite familiar to us today in the fields and forests around us. The empire was heavily reliant on spelt wheat to feed its masses, which had a ripple effect on the make-up of wildlife in Britain. In fact, this swelling of the land with crop-filled fields seems to have resulted in a major shift for biodiversity in Britain. Although, as we will see below, there was a great deal of new plant and animal life brought here during the Roman period, the empire's exploitation of Britain's land to provide grain resulted in a marked decline in pollen diversity in the archaeological record. You see, the agriculture of prehistory had actually resulted in an overall biodiversity *gain* – the mosaic habitat of mixed arable crops combined with the weeds that hitched a ride with them, like stinking chamomile, henbane and poppy, all helped to increase the abundance of different pollens now trapped in the layers underground. Ecologically speaking, the time around the Iron Age-to-Roman transition boasted the richest landscape we've ever had. But at some point the insatiable appetite for land, resources and standardisation began to have a negative effect. For those bits of Britain within the empire, Roman occupation was the final nail in the coffin for this brief harmony – pollen analysis heavily suggests that the countryside became more homogenous.[7] In effect, this was a landscape that was now falling victim to the market economy. Land was no longer just what we lived in and on but came to represent something far more abstract. It was, for the first time, a commodity.

Selling nature

Sometime in the years of Hadrian's reign in 117–138, a surveyor rode out to an area around Canterbury to examine a woodland. This five-acre patch named Verlucionium belonged to a Roman citizen named Lucius Julius Bellicus, who had purchased it from Titus Valerius Silvinus for forty silver coins (*denarii*). The surveyor established the boundary of the wood to be between land belonging to two others: on one side to the heirs of Caesennius Vitalis, on the other to those whose names have been lost in time. All of this information is recorded on the first page of a legal document etched onto wax on a tablet of silver fir just like the Bloomberg Tablets. In 1986 it was recovered by archaeologists from within a pile of refuse near the now-subterranean River Walbrook in London. As with other tablets the wax had long since disappeared in the wet conditions, but part of the surveyor's story was preserved. Given the wording on the tablet, that Bellicus was 'testifying', and even that there was a surveyor in attendance at all, indicates there was a dispute over this patch of land. It has been tentatively suggested that the answer as to the nature of the argument may lie in the woodland's name, Verlucionium, thought to have pre-Roman origins and meaning 'light' or 'shining grove'. Sacred groves, including those that pre-dated Roman rule, were supposedly automatically made public land in the empire, which by extension meant that they could not be bought or sold. Closer to Rome, woodlands were important sources of timber for villas and vineyards but were often detached and at some distance. As such, there was regular illicit private trade in public woodland, which prompted the need for inspection and record-keeping to protect them.[8] Either way, the tablet represents perhaps Britain's earliest legal document pertaining to land.

Elsewhere, the privatisation of land is sometimes hinted at in the archaeology, such as at a villa in the Yorkshire Wolds, at Rudston. This sits in a landscape that in the Iron Age was dominated by long, linear ditches, known as dykes, enclosing large areas that were farmed communally. In the second century AD, however, they were reorganised from large areas into individual strips, indicating that private ownership was replacing a communal regime. A century later, a villa was constructed on the site, perhaps the final bit of local adoption of Roman culture.[9] With privatisation came the opportunity for land to be exploited by absentee landowners, who were able to reap the rewards from land worked by tenants and slaves. This was a pivotal moment in the history of our landscape – it set in motion ever-increasing control over the land and what happened to it, and is not something from which we ever really recovered. Even though not every Roman field, village or town survived into the present, the notion of land rights as something to be traded between individuals certainly has. It has been passed through several filters in the intervening years, but the fundamentals are still there. Space can be reserved exclusively for some at the expense of all others.

Origin stories

Clearly, conveyancing is not the most memorable or oft-cited gift the Romans famously gave us. That title sometimes goes to things that occur 'in the wild', among them the sweet chestnut tree, the opium poppy, apples (other than crab apples) and oysters. *Some* of these purported introductions are even true, but many are just myths that persist because the idea of 'being Roman' has such a strong draw. The nostalgia and even romanticisation of Roman occupation and empire in the eighteenth century, when many plants were being formally classified and 'official' narratives were created, drove a desire to connect things with this glorified age.

First, then, it is probably wise to debunk some of these myths, and we'll start with the sweet chestnut. There is no evidence for the tree earlier than the seventh century in Britain, and that is only a single occurrence in the archaeological record before the thirteenth century. The first written evidence dates to 1113. Its own origin myth appears to have come largely from John Evelyn's *Sylva,* in the late seventeenth century, in which he speculated that it wasn't native and that Pliny's instructions for using sweet chestnut for vineyard stakes meant it had probably come here with the Romans. Since then, fanciful writing has suggested that whole groves of sweet chestnuts were grown for the use of Roman armies, but we simply don't have the evidence. The trees we have now are genetically consistent with introduction from southern France, northern Iberia or northern Italy, as they do not match those found in Turkey and Greece, but the oldest that can be definitively dated appear to be only as old as Evelyn's *Sylva*, despite the fact that sweet chestnuts are often held aloft as our oldest trees.[10] Another myth relates to oysters, which are a bit different in that they aren't an introduction at all: they are native but were much more heavily exploited by Romano-British communities than in prehistory, and so appeared in the archaeological record as a new intrusion. The opium poppy was also certainly here in the Roman period but has appeared in earlier contexts so cannot have been introduced by them.

So, what of those that *are* Roman introductions? Perhaps some of them will come as a surprise, those that have become emblems of the British landscape, gardens, parks and 'wild spaces' alike. The 'English' elm, not the elm that had come with the warming winds of the Mesolithic but another that has been so stunted by Dutch elm disease, is not 'English' at all. It originated much further south and is possibly a Roman transplant. *Helix pomatia*, commonly known as the Roman snail, was introduced for food. These large, ridged snails have a natural range across a central band of Europe, and are still famously popular in French cuisine. Damsons, now so ubiquitous in the hedgerows, were transplanted here, likely from the Eastern Mediterranean, and then spread across the Roman world. Fruits like the peach absolutely were consumed in Britain, but the evidence is not clear enough to say they were grown here until later on; the evidence we do find, restricted to towns, must be from imported fruit.

For so many others – asparagus, parsnip, walnut, sycamore, oilseed rape, cucumber and coriander, for example – the jury is out. It's not possible to say whether or not they were an established presence prior to this period and just cultivated with more intensity so that they show up more readily in Roman samples. Sadly, we may never know the origin story of the humble carrot.

The first gardens

Many of the new plants were not really intended to self-seed into the wild but were grown in the first gardens. We have seen in the preceding chapter that there was small-plot cultivation in prehistory in what we might think of now as gardens. But what makes a garden a garden, rather than a small farm? What sets the Roman gardens apart from small defined areas under cultivation in prehistory? What makes it a garden most of all is something that spilled out beyond the confines of a plot and into the landscape beyond – cultivation and construction for aesthetics as much as produce. We know from literature, wall paintings and archaeology that Roman gardens on the continent were highly organised and decorative affairs. They had gravel paths, pillars and fountains; fruit, vegetables and herbs; flowers for the bees and pergolas for shade.

Garden archaeology can be very tricky to spot because plants are ephemeral and can disappear as quickly as they're planted. Because so much of what makes a garden is above ground, there may not be much to find in the ground when the main components have rotted away. However, it's not impossible, and in the last few decades there have been increasing numbers of excavations aimed at investigating what may have been grown. They are invariably targeted at areas where we know from documentary evidence that there was a garden, rather than being found by chance. There aren't many, but there are some places where a Roman garden has been inferred from the archaeology. At Gorhambury near St Albans, a first-century villa was built with an avenue of trees or hedges, and possibly a structure for training climbing plants. This might sound a stretch, but we have some ways of knowing. In this case, large pits with fine loam soil were excavated, *so* fine and stone-free in comparison to the surrounding soil that the excavators believed it could only have been sieved into place. These were regularly spaced and in a line, and would therefore seem to have been the kind of planting pits that we would be familiar with today, for shrubs or trees. Similarly, regularly spaced post holes indicate that this avenue also had an arbour.[11]

Sometimes plant remains are revealed in contexts such as burials. Several burials and cremations, particularly in waterlogged deposits where plant preservation is better, have yielded leaves and twigs from what appear to have been wreaths made of box. Box is an evergreen shrub, a mainstay of gardens now and well known for its topiary. The discovery of leaves that appear to have been cut rather than fallen has been taken as evidence of topiary in Roman Britain, and Fishbourne Roman Palace may have had a box hedge. Rarely found but also making an appearance on some sites are other evergreens like stone pine and, perhaps surprisingly, Norway spruce, our classic Christmas tree. These were utilised for their aesthetic and their architectural quality in the gardens of the time. Britain had its own evergreens, such as holly and juniper, that seem to have carried with them deeply held meanings. The yew is considered one of our most folkloric trees, and perhaps that is why it doesn't seem to have been used in topiary until after the medieval period.

Rather than focusing on whether something was *introduced* in these few hundred years of Roman presence, it's probably more helpful to think about the categories of living things that evolved. Prehistory primarily gave us animals and plants that we kept and controlled, moulded to fulfil particular duties that were largely practical – food, materials for warmth and clothing. With the Roman presence, new categories of animals and plants became visible in the ways people went about their lives, as a natural consequence of the way in which

Roman society and its hierarchies played out on a grand scale. Moving beyond animals as food, labour or as mediators of the supernatural, these new categories included pests and pets, animals for entertainment and those kept specifically because they appeared exotic.

Pets and pests

Along with the benefits some introductions brought, Britain also suffered with new and voracious mini-predators that hitched a ride, accidentally exploiting the massive trade network that connected so much of the European mainland and North Africa to Britain. There are very few animals with global distribution, but some of them got their foot in the door with the Romans. The perennial pests within the baking drawer, those tiny little bugs that find their way into the flour, are one of a number of grain beetles that became established simply because of the sheer volume of food that was required around the empire. They arrived with the earliest of the Roman soldiers and are present on urban, fort and rural sites, though the highest concentrations of them seem to have been in lower-quality grain that was used as animal feed – they show up in manure. Some of these kinds of insects may have been present in Britain already but it was the sheer scale of farming and the number of people it served that meant they thrived to the extent they are so ubiquitous on archaeological sites from this period.[12] Although it was the Victorians who had a real knack for introducing plants that became pests, the Romans did it too. Ground elder, for instance, was cultivated here as a food crop, as its young leaves are a tasty addition to salads and other dishes. However, it's an absolutely unstoppable little plant that spreads via rhizomes, meaning that if you chop it up or even try to dig it up you only succeed in making it multiply. I've largely given up trying to control it in my own garden and have resorted now to feeding it to our guinea pigs, who love it, which is some small comfort as I watch it drowning out the things I'd actually rather see growing there.

A more nefarious character also made its British debut, scurrying off ships and gorging on stored grain, carrying with it the pestilence that would go on to decimate populations around the world more than once: the black (or ship) rat. This rat is not the same as the one we are most familiar with today – especially if you live in the city, near or on a farm, or keep any kind of animal outside. *That* rat is the brown rat, also known as the Norway rat, and we'll return to its story later. The black rat is smaller, with bigger ears and an even longer tail. Originating from South Asia, it was to become a species fully commensal with humans: it came to rely on us and our way of life entirely for its sustenance, survival and spread. By the early first millennium BC the black rat had moved with people to the Eastern Mediterranean, and then hitched a ride with the expanding empire. It looks like the rat landed here with the earliest Roman ships, or at least that is when it becomes common enough to be archaeologically visible. A rat was found in charred grain that was preserved when a store burned down in London during the Boudiccan revolt in AD 60–61. Much of the grain itself seems to have been imported from southern Europe or the Mediterranean as this was prior to the establishment of heavily monocultural crops in Britain, and included wheat mixed with arable weeds, such as the earliest example of corn buttercup known here. It is possible the rat itself arrived with the same shipment.

It's difficult to overstate the impact that the black rat has had on world history, yet its population has not been uniformly strong across the board, perhaps mirroring the fortunes and intensity of the human populations on which it relied. For most of the roughly four centuries when Britain was part of the Roman Empire, black rats seem to have been confined to the cities. Towards the end of the fourth century, however, they started to appear in rural sites, such as a small villa in Oxfordshire that was abandoned at this time. Centuries later, the careful sieving of sediments on the villa excavation revealed the tiny mandibles of several juvenile black rats, along with other small mammal bones, having passed through the digestive systems of barn owls.[13] We know that, as human activity at the building dwindled, an enterprising barn owl moved in and reared its young for at least two summers in the rafters of one room of the old villa, while the roof, or at least a portion of it in the corner of one room, remained intact. These summers of occupation are likely to have been after the last people had moved on, as the pellets they dropped congregated in an area roughly the size of a double bed and were not cleared away. By 385 both the owls and the rats had moved on, no longer able to survive on remnants of food left by the people who had lived there. This site actually mirrors a broader trend in Roman archaeology – black rats seem to have largely disappeared from Britain for the whole of the early medieval period, from the fifth to the tenth/eleventh century, and it was for this reason that for a long time they were thought to be Norman introductions. It might be that the species simply spread into the rural population too late in the day to take much of a foothold outside of the cities, some of which were abandoned. They hadn't made it to Scotland or Ireland, and only one example is known from Roman Wales.

Another animal that came with and perhaps *because* of the black rat was the domestic cat. Beloved by many but treated as a pest by others, there are few animals so ubiquitous in our daily lives and so divisive as the cat. They enjoy a curious place in our culture, castigated for their impact on wildlife while simultaneously employed to control rodent populations. We have good reason to think seriously about the significance of free-ranging cat populations: their imprint is tangible beyond their predation of birds and small mammals. They take prey from other wild animals, such as hawks, and their scent and movement disrupts animal behaviour and habitats. Their ability to hybridise with wildcats also seriously threatens some indigenous species. Really, we could classify the cat as an invasive species – there are very few places free from domestic or feral cats. One study estimated that cats in the UK brought home 57 million mammals,

27 million birds and 5 million reptiles and amphibians over a single five-month period.[14] Even using the conservative estimate that cats bring home 25 per cent of their catches (some suggest it's closer to only 10 per cent), scaling those figures up would indicate that cats in the UK wipe out approximately 547 million mammals, 259 million birds and 48 million frogs and reptiles each year.

There is some evidence for the cat in Iron Age Britain; perhaps it had snuck off boats or even been given as part of trade deals in earlier times. These early cats were very similar to those we have today, without the diversity brought about by selective breeding in the last couple of centuries. The ginger cat, however, seems not to have spread out from Asia Minor until the post-Roman period. As with the brown hare, there is a sharp rise in the numbers of cats found on archaeological sites in the Roman period. Cats even made it as far north as Orkney, despite the old adage that the Romans didn't make it that far. It has previously been argued that these finds either represented wildcat, despite the skeletons being rather too small, or there was a larger population of domestic cats already in Britain, with wider distribution. Literary sources describe the Roman circumnavigation of the British coast and some thought they could have been accidentally introduced during these trips. Genetics have now shown that the Roman cats found in southern Britain share the same DNA as those in Orkney, indicating a single period of introduction from one source. They may well have been favoured for their role in helping to keep down the number of rodent pests that were inherent to the trade network, but undoubtedly were also kept as what we would consider pets and enjoyed a status beyond the utilitarian. Pet-keeping more generally was on the rise: caged birds were popular in elite households, and dogs began to take on a different form. Almost all of the dogs in Iron Age Britain were of medium to large size, but now came many much smaller dogs, including those with 'dwarfism', similar to modern dachshunds, and lapdogs, which must have been the result of deliberate human selection. There are skeletons of chihuahua size and those similar to Maltese or Pomeranians which were imported here, though the more mastiff-sized hunting and guarding hounds were exported to the continent.

The wild hunt

We know that people of later prehistory were not relying on wild animals for food, and that if they were hunted it was not with any great frequency. Wild animals had become, possibly, taboo creatures to bring back into the home. But with the Romans came an entirely new set of ideals, and with them, a new set of wild things that were introduced as hunt prey. It wasn't that animals were released and then immediately able to thrive; the process was more protracted. This in itself indicates how tightly controlled some animals were, in terms of both who had access to them and how they were kept in the first place. Later characters like the muntjac deer have assimilated into the British countryside much more quickly, and this cannot just be because they are better suited to it. Nonetheless, there are dependable beings in our countryside who owe their place as icons of the rural landscape to the Romans.

These animals tend to be most visible on elite sites, probably because most of them were hunted for entertainment or status and this was confined to the upper classes. The Romans didn't hunt for food, particularly, but do seem to have been happier to eat wild animals than people were in earlier periods, and there is more evidence of wildfowl and game in the record. Most sustenance was from farming or imported food, and sometimes even when things *could* be grown here, they weren't. Although we know figs were being eaten we don't think they were cultivated, even though the climate is estimated to have been 1–2°C degrees warmer than today. Members of the Roman army hunted in their spare time and chose animals that were seen as challenges rather than those that would feed them. One commanding officer of a cavalry unit in the area around Binchester in

County Durham dedicated an altar to Silvanus, the god of uncultivated or wild land, after he had managed to best 'a wild boar of remarkable fineness' that others had failed to kill. For the first time it seems that the *challenge* of the hunt became more important than the consumption of the end product. Hunting had become a sport.

To feed the hunger for good game we see the first enclosures to keep wild things *in* rather than *out*. In these proto-parks, animals such as fallow deer, hare and rabbit were all kept as part of a new style of entertainment. Both fallow deer and rabbit were, like the black rat above, once thought to have been Norman introductions, but we know now, through advances in scientific dating and a reassessment of archaeological material from museum stores, that they were present in the Roman period. Like the black rat, they do not seem to have survived the demise of the empire – they are absent from the archaeological record from the fifth century until after the Norman Conquest. Although today rabbits are numerous across the countryside and are well known for their fecundity, early rabbits were not hardy creatures and are likely to have struggled in our climate. In fact, they didn't become naturalised here until around 300 years ago, despite being bred in huge numbers in rabbit warrens from the thirteenth century onwards. They hail from Iberia, and burrowing is part of their protection from the cold as well as from predators. Unlike the hare, the rabbit is born hairless and blind, and the medieval keepers at least provided them with their very own bespoke and artificial burrows. It is likely that Roman keepers also provided them with some shelter. Given this, it seems that if any were left when these things fell out of fashion, they were unable to colonise the countryside. Rabbits were probably always reasonably few in number because they were high status and ownership was restricted, and those we do find are often shown to be intrusive – their habit of burrowing means they have physically found their way into archaeological layers but date from much later times. The failure of fallow deer to thrive after the Roman Empire shrank is more confusing, but certainly the genetic evidence indicates that they became extinct here for several centuries.

At Fishbourne Roman Palace, so named because of its vast size covering over 5.5ac, a very high number of fallow deer and hare bones have been found which hint at the existence of *leporaria* (enclosure for hares) and *vivaria* (enclosure for deer). In these enclosures, game animals were bred, displayed and hunted. Fallow deer were probably brought here because they were better suited to being kept in enclosures than either the indigenous red or roe deer. Although Fishbourne is unusual because it represents a very large elite complex, it does seem to be a good proxy for wider trends that were occurring in contrast to the time before. Across high-status sites are good numbers of sheep, cattle and pig, but also much higher numbers of wild game, wildfowl, fish and wild plants. One of the fallow deer at Fishbourne was shown through stable isotope analysis to have been imported to Britain as a fawn, early in the occupation. It was as if the Romans had moved into a new house and set about redecorating with fervour.

Although the hare was not a new introduction and small numbers were present in the Iron Age, their numbers rose. Caesar remarked in his accounts that the Iron Age Britons ate neither hare nor chicken, and that's supported by excavation evidence where they have been found to have been buried intact. This trend seems to have continued. Instances of iconographic representation go up considerably over the course of the Roman period so perhaps depicting them was no longer taboo, but there are particular clusters of objects in places like Norfolk. This was part of the territory of the Iceni when the Romans invaded, home to Boudicca, who is famously said to have released a hare from her gown before battle with the Romans. It is thought that hares already had a special place in Roman culture and that this was not borrowed from incumbent cultures. Hares were also high status in continental cultures, so with fresh imports also came new deities that seem to have morphed together in some areas, associated with the Celtic Andraste, goddess of war, as well as the Roman

Diana, a goddess of the wild and the hunt. In the post-Roman world, the hare faded into the background for a while and is rarely uncovered from digs. I'm afraid to say at this point that although the Easter bunny is often said to have been a hare and connected to worship of the goddess Ēostre, it's not true. A recent study has demonstrated that while there may have been a deity known as Ēostre, her worship was localised to eastern Kent and there is absolutely no evidence she was ever associated with hares, or rabbits. In fact, this association seems to go back no further than the seventeenth century, with the Easter bunny then coming into being in the Victorian period.[15]

Wild trade

The new animal inhabitants of the land may have begun as truly wild in their home countries but their place in Roman Britain was frequently as imported exotics to act as hunting entertainment in kinds of menageries. It's difficult for us to think of them as exotic because they are woven into the fabric of what we understand to be our British wild. Hares were seeded here, becoming another part of the backdrop, like the damsons we forage from autumn hedgerows.

The fact that these animals were introduced and kept in their own menageries highlights that just as the land itself became a commodity and a tool to be wielded by the elite, now wild animals came to be owned. Like their domestic counterparts, they were able to become property by virtue of occupying land that was seen as belonging to someone. Their value lay not in their tameness or taste, but precisely in their *wildness*. Hunts involving wild boar, hare and deer are depicted on mosaics and cups in such frequency that to hunt was clearly performative. In this way it's almost as if hunting was for other people, not for personal consumption or enjoyment. The parks that some exotica were kept in were designed to *look* natural, just like we will see again in the landscape park trend of the eighteenth century. Roman writer Columella even suggested including woodland in villa estates, perhaps specifically to keep deer, and I wonder if this was the aim of the woodland transaction described above. It was not unique to the Romans – in China, in the first millennium BC, landscapes and parks were designed in exacting detail, all to look like they had occurred naturally.[16]

Away from the natural aesthetics of estate entertainment, gladiators were pitted against the most exotic and ferocious creatures in arenas called *venationes*. Sometimes, the beasts were dressed in costume for the entertainment of the masses. There flourished a trade in exotic animals like tigers which were shipped in crates from ports around the empire. There is a mosaic in Ostia, Italy, showing that wild animals could be ordered from the port, and they may have been caught to order. Pliny even described one sculptor who was eaten by an escaped panther on the docks while he looked for real animals to act as models for his work. We don't have huge amounts of evidence for arenas in Britain, but it was in the decades after invasion that the Colosseum in Rome was built to house such entertainment, with room for 50,000 people to watch 9,000 animals being slaughtered during the opening festivities. It does seem likely that similar events occurred across the Roman world, including here in Britain. These sorts of animals are also visible in a group of mosaics depicting Orpheus and his lyre. Most commonly depicting lions and leopards, there are several examples dating to the fourth century in Britain. Orpheus was a figure in Greek mythology who was given a lyre by Apollo, the god of music, among other things. The lyre enabled Orpheus to charm wild beasts and birds, who surround him as he sits, propped up by a tree, making music. In Britain, such mosaics are found in the reception areas of villas, and would have projected the status of the owner, with the power of Rome behind them. The use and display of such animals had the twin effect of demonstrating Rome's vast reach and ability to access far-flung lands with unusual inhabitants, but also spoke to the power it had to nullify such beasts. The civilising power of Rome could charm the beasts and placate the wilderness. The wild was at once something to fear, something to be bested, and something to be controlled. This is an attitude that was not so obvious again until the medieval period.

Eco-anxiety

Beyond ownership and control, the practice of enclosing animals in parks as the privilege of one group of people belies another attitude – that is, that the Romans understood wild things to be a *finite* resource. They needed to protect the wild for the people they supposed deserved access to it the most, but were also anxious about the changing landscapes and feared that the gods would punish those who killed young animals or did undue harm to the world. This was complicated by the attitude that anything beyond Rome that was wilderness *should* be civilised and tamed. Because of the growing role of animal entertainment in the late Roman Republic, people had already witnessed the extinction of some animals and the curtailment of others' territory. The range of animals like lions in places such as Greece and Italy had become restricted to the mountains. In the first century AD Pliny wrote of the loss of a popular plant called silphium, which seems to have been a Mediterranean fennel-like edible with a long stem and clusters of yellow flowers. It was difficult to cultivate, with efforts to do so resulting in a plant that was not as potent as in its wild form. Pliny believed the last stalk to grow of its own volition had been worth its weight in gold and sent to Nero as a gift. It wasn't just that it was gone, but Pliny specifically bemoaned the role that people had had in its extirpation – farmers had stripped the pasture with sheep but it had also been generally over-exploited.[17] We can't really be sure that silphium was or is extinct because we're not entirely sure what plant the term refers to, but the pertinence lies in the fact that it was recognised as a casualty of human desire. It also shows something of the attitude, perhaps snobbery, on display elsewhere in Pliny's writing, that cultivated forms were never quite as good as wild ones. It all sounds rather familiar. Other authors of the ancient world also hinted at a certain nostalgia for the way things used to be done. Theophrastus, writing in the third century BC, had acknowledged not just broad weather patterns but changes in the microclimate that were the result of deforestation and human-directed changes to watercourses. We cannot say that this is unique to the Romans – we know that other animals had been hunted to extinction in times before, and have seen how some believe ecological grief was palpable in the veneration of the oak at Seahenge. But the *pace* and *rate* of change accelerated in the Roman world, as things became ever more organised and fixed in place.

Tamed

Let's return to our 'exotic' and chimeric wolf mosaic, which delights in its comedy and horrifies in its art. Perhaps it's unfair to criticise an artist undertaking a feat few of us would be successful in, but one does have to wonder exactly what information they were working from. If the artist had never seen a wolf, it's difficult to imagine anyone telling them it looked anything like a horse, especially when there were so many dogs around to compare it to. The mouth could be interpreted as being in the shape of a smile, but by bearing all its decidedly pointy teeth it is more likely intended to be the snarl a she-wolf is often depicted with, representing her wildness. It could simply be a case of an inexperienced artist – a browse through other inscriptions and reliefs found in Britain shows other poorly illustrated animals such as wild boar, sometimes looking more like dogs, and dolphins, appearing more akin to evil sea serpents. But boar, too, were already uncommon in Britain at this stage, and how many of us have actually seen dolphins in the flesh? Even if there *were* wolves in the countryside or being imported into the arenas, the vast majority of people were unlikely to have really encountered them.

Taking into account the rural landscape that the Romans invaded, already denuded of so much of that which we consider to be 'wilderness', and coupling that with all that Rome brought with it – gardens, arenas, cities, vineyards, intensive mining – it is very difficult to imagine that anyone was encountering animals like wolves with any frequency, even if they were present. We almost need to divest ourselves of the image of the past as a space in which people frequently chanced upon and spent time with wild animals, like some kind of character from a mid-twentieth-century Disney cartoon. The creation of permanent homes for people in ever-growing settlements took habitats away from animals and plants, squashing them into their own zones. There's a very good chance the wildest things people came across in the later years of Roman rule were in the form of curated artworks rather than actual wild animals. These were images on moulded and decorated pottery vessels, wall paintings or mosaics. The richest and most well connected will have been exposed to exotic animals from abroad in the flesh, but we know that most people in Britain didn't enjoy that kind of lifestyle. The cultural transmission of figures that dealt with the wild, such as Orpheus and Diana, ensured that any encounter with these spaces and their inhabitants was already imbued with something altogether more intangible. Forests and marshes were the homes of nymphs and sprites, but although some writers point to a fear of the wild, it seems to be the people who dwelt within what they considered to be the wild that were the thing to be feared. Tacitus told of the barbaric tribes in the north whose lack of civilisation was the thing to be feared, rather than their aggression. After all, Rome had an organised army, with marching legions covered in armour. So, although things were valued for their wildness, perhaps they were not feared, except by those who necessarily came into close contact with them. There was some balance offered by the gods that had to be appeased – the Romans were controlling, but they weren't entirely superior.

The occupation was really a time between two worlds. While some Celtic gods had been rebranded – Sulis to Minerva at the famed baths of Bath, for instance – others survived more or less intact, such as

Epona, a horse goddess known from Iron Age Ireland and Wales. But when Christianity – a religion that was to really cement the growing perception that the nature of the world was hierarchical, and we were above all others – took root in parts of England during the third century, it heralded the end of polytheism and of divine beings that would place anything but themselves above the human. This was a religion in which the natural world was created *for* humans. This served to exacerbate the feeling that what was outside of Rome and outside of Christendom was a wasteland of wilderness to be braved and brought to heel.

Roman retreat

Nothing lasts forever, so they say, and a whole host of factors came into play when the imperial administration eventually disappeared from Britain. There had been raids with increasing frequency from groups including the Saxons, from the east, and Picts, beyond Hadrian's Wall. Everything had become a bit anarchic. Having not been paid for several years, the remaining soldiers nominated their own emperor, who then left with most of them to wrest control of what was by then the Western Roman Empire – half of the whole that had been divided between Emperor Theodosius' sons in 395. This was not a desertion, and the Romans didn't really leave; the vast majority of Britain's inhabitants remained here. It's simply that the governing structure had crumbled, and many of the things that made up Roman identity and upon which it relied had begun to disappear. Gone was the mass manufacture of Romanised pottery, and coins were no longer minted. Urban centres declined, and much engineering was lost.

We don't know where the family who prospered under Roman rule went when they left Althorp, whether they moved on when it became too difficult to maintain their way of life without access to everything the empire had provided. Sites like this often seem to have been closed down deliberately rather than slowly abandoned, and the demolition layers we dug through to get to any intact building foundations certainly suggests this was the case. Much of the building stone was thrown into a well, along with coins, and I imagine many other things, but sadly it was too deep to excavate safely.

The occupation of Britain was a story of boom and bust in a way, but the landscape was always a lived one and it wasn't an entirely homogenous experience. It was diverse in its inhabitants and ecologies, despite the growing sameness of its agriculture and other imperial infrastructural additions. But, it would seem, by roughly the second century a major biodiversity shift had occurred, and the land ceased to be 'wilderness'. There were very few, if any, areas where there was no Roman settlement, except in the most upland of areas, like the Pennines. Some of the famous road network linking disparate parts of the land also survived to an extent. Any straight section of road now has me wondering out loud if it's Roman, and often it can be traced by zooming out on a map and looking for the continuation of the line somewhere, even if that's some distance away. Field configurations can also reveal the routes of old Roman roads even when the route (or some of it) has disappeared. When the occupation was over and the use of the landscape was less intensive, some habitats made a bit of a comeback and sometimes this regeneration is demonstrated in their names. Those with derivations from *wood*, which is early medieval, but that have extensive Roman settlement indicate an area that was deforested during Roman occupation and then recovered enough to be named as a wood. This does not mean that the same kind of wood replaced it. In France, for example, present-day woodland biodiversity has been shown to correlate strongly to areas of Roman cultivation and settlement. That is to say, even with intervening *over*-exploitation of the woods in the post-medieval era, what is there today still echoes human activity that occurred nearly 2,000 years ago. It doesn't seem likely that this is because the plants growing in areas of Roman cultivation are descended directly from those planted back then, but rather that an area's character is informed by the impact ancient activity had on the soil composition.[18] It really does make me wonder more about the forest of Białowieża and our insistence on considering it a primeval wonder; I don't think anywhere we've lived can really claim that title.

4

CONTROL

(MEDIEVAL, 410–1540)

Hunting status

> Jack Wade was never so afeared as when the hare trod on his beard.[1]

So goes a fifteenth-century Scottish poem describing a farcical hunt. A yeoman riding out has spotted a hare and calls on the gentlemen of the local village to provide greyhounds to join him in the chase. Upon informing the yeoman there are no gentlemen there, the local steward swiftly reels off a list of all the village men who have dogs. Being landless commoners, however, these people were not permitted to own greyhounds and so instead arrive with a motley bunch of animal-baiting mongrels. The yeoman, disappointed but undeterred, springs the hare by startling it with an arrow. Instead of running away, the hare rushes towards and attacks the crowd of ill-prepared people and dogs, who, having grown restless, are already attacking each other. The hare eventually makes it to safety, but not before inflicting a great many injuries on the men, who must be collected and taken home in wheelbarrows by their wives. This poem is of course a comedy, but it is also a comment on the status of the people and animals involved. Following a drop-off in the status of wild animals at the beginning of the early medieval period, hares had risen in stature once again to be considered a noble quarry that deserved only the finest and most considered type of hunting – they were to be coursed with elegant sighthounds who were reserved by law for people of a certain standing and income. The wild and clever hare in the poem outsmarts the peasant hunters who are attempting to bait it like they would a cockerel or bear at the fair. It is an allegory of the deeply rooted, class-based access to hunting and wild things that had carried over from the Roman period and become entrenched by this time, but also carries something of the magic that was now seen to be imbued in the wild.

We tend to speak of the long medieval period as one of two halves: pre- and post-Norman, or early and later medieval. The archaeological record is sparser after Britain ceased to be part of the Roman Empire and historically this period was often referred to as the Dark Ages, mainly because of the lack of written sources. We do also find literal dark sites – places where there is a lens of very dark humic earth covering old Roman layers. Previously interpreted as abandonment phases, these places are now understood as representing a combination of agriculture and waste deposition, as well as abandonment, from those following a different lifestyle more dependent on food production than trade, in what had been Roman settlements. The pre-Norman years, covering the fifth to eleventh centuries, is also often referred to as the Anglo-Saxon period, after the people who settled in eastern and southern England from elsewhere in northern Europe. The Saxons were just one of the cultural identities that existed in Britain at that time, alongside Picts in Scotland, Gaels in Ireland and migrations into all parts of Britain from, most notably, Scandinavia and Frankish territories (encompassing much of modern-day France, Belgium and parts of the Netherlands and Germany). Differences were expressed through clothing, language, jewellery and even burial practices, and although there is some continuity from Roman rule, the archaeology tends to indicate a reversion to older ways and earlier attitudes to wild things, for a time at least. Eating wild animals such as boar was possibly taboo, and it is not until the tenth century, shortly before the Norman invasion, that they start to reappear as food waste.

Surviving fairy tales that tell us of witches, wolves and woodsmen are largely responsible for the popular view that the countryside of medieval Britain was 'wild', but it is also a reputation that grew out of contrasting it to what had come before. The Romans have been described by history as the ultimate civilising force, and so the periods that came both before and after have been seen as far wilder. However, even though the Roman period had seen biodiversity begin to decline, the expansion of agriculture and introduction of private land hadn't yet wiped the wild away, and it is the medieval period that really saw the total domination of the landscape of Britain. Areas were designed and set aside for leisure, gardens for growing native and imported fruit, vegetables and flowers, and the first commercial fishkeeping. Fish rose to be of particular importance as a proxy for meat, because in the Christian world eating meat was prohibited on Fridays. At Clarendon, near Salisbury, the ruins of a medieval royal palace survive, with an inner park from which long deer-coursing matches could be watched, and an outer park boundary that made it the largest deer park in the country. The palace was essentially a country retreat deep in the forest, but historical accounts, maps and archaeology reveal this was also a heavily managed setting where every bit of nature was working in some way for the monarch.

Across Britain, there were perhaps vestiges of ancient woods surviving in small pockets – place names deriving from old Germanic terminology for woods, such as *wald*, *leah* and *wudu*, concentrated in south-east England, and the Viking equivalent, *thwaite*, in other areas, point to tracts of woodland fossilised in memory if not reality. But though these places *may* have held the lingering remnants of the *wild*wood, by the end of the medieval period they had become such managed spaces that they had only the weakest claim to that prefix. Nothing was truly free to roam any more, whether animal, plant or human. After the conquest of 1066, the Normans, like the Romans a millennium before them, imported their own ideals, animals and plants to Britain. The most enduring of these imports were not building designs, trade networks or pottery styles, but laws.

What ran through this period was a desire for control, of rights to use land and animals, and of using these to control each other and uphold social hierarchies. The wild as an entity became entwined with status, reflected in the stories told about it. Ordinary people may have relied on gathering herbs and plants for everyday use, collecting wood for mending cottages and feeding fires for warmth and food, but these activities were increasingly constrained, some taking on new meaning as the stuff of darkness and witchcraft. Cautionary tales emerged, the literary ancestors of stories such as 'Hansel and Gretel' and 'Red Riding Hood', that warned against wandering in the wilderness. The people who lived in the woods in these stories were suspect characters, witches or cannibals.

We continued to clear woodland, diverted rivers and streams for mills, drained land for cultivation and covered the most fertile bits with fields organised into 'strips', evidence of which is fossilised in the curving banks and ditches present in so many pasture fields today. Hunted animals like wild boar and deer had new status, and the rich consumed them conspicuously, while the poor were just as concerned with keeping wild animals away from their own small stock of domesticates. In manipulating habitats for animals like deer, hares and rabbits, the rich preserved their quarry, but escapees could decimate a small-farm crop. True wilderness, such as the oak wood and birch in the far north perhaps, was dwindling, though a few wild animals clung on where they could, before disappearing into extinction. The Domesday survey records that only around half of England had any woodland by 1086 and, as British historian and ecologist Oliver Rackham once wrote, eleventh-century England would have looked more like modern France than Borneo.[2] The medieval period was the adolescence of the landscape we know now, when the landscape of Britain became wholly domesticated.

Gardening Eden

Much in the medieval countryside would be recognisable to us today, particularly following the Norman Conquest. There are tangible legacies in churches, town layouts and quaint timber-framed jetties overhanging old buildings. What people may previously have done 'in the wild' was often brought into the realm of the domestic and given dedicated space – fish were given ponds, and many more plants aside from crops were deliberately grown rather than foraged. There were still common experiences in the forests and fields, which we'll look at below, but increasingly areas became more bespoke and more individualised.

Nearly 90 per cent of people in Britain today have access to either a private or shared garden space, which we use for trampolines or climbing frames, flowers and patios. The majority will feature neatly clipped grassy lawn areas or even faux versions made of plastic grass so as to maintain the appearance of greenery without the hassle of mowing. In the later medieval period, gardens were common, but they certainly weren't normally or predominantly lawns, and where they did exist they often carried far more meaning and were not just there to look good. Green had become a revered colour, that of chivalry and spirituality, but very closely cut grass also encouraged tiny wildflowers to take root. Gardens tended to range from back plots used as small-scale farms on which a few animals were kept and vegetables and fruit trees grown for individual use, to highly designed, segregated and walled enclosures with a variety of turves, herbs, flowers, trees and water features. Not all were for sustenance but in monasteries and noble houses alike there were garden spaces set aside for quiet contemplation as well as for growing plants for medicinal use, living designs in trained roses or yews, or vineyards. The idea of an aesthetic garden was not completely novel but it certainly spread more widely, and took on a wealth of new influences. From the eleventh century much was drawn from Islamic paradise gardens (the idea for which arrived in Britain via Norman links with Islamic culture in southern Europe) – in fact, it is thought that the word 'paradise' comes from the Persian word *pardis,* used for these beautiful gardens behind walls. They were often arranged into four parts divided by running water, and offered tranquil, sensory and safe spaces that were lush and green in contrast to the arid – and conceptually wild – desert environment beyond. They are the forebears of the medieval *hortus conclusus* that became commonplace in Christian art, complete with depictions of the Virgin Mary. Like the paradise gardens, these organised areas were enclosed by hedges and sometimes walls, and acted as a haven from the wild world beyond, even though now the wilderness, in Britain at least, was beginning to take on a far more fictitious existence.

Beyond the wall

The power of this perceived wilderness weighed heavily on the medieval mind. Many references to the wild are imaginary, used as something with which to juxtapose the civility, safety and acceptability of Christianised space. It's not very different to how we think about it today: somewhere we long to escape to, but where we may need our wits about us.

The idea of wilderness as lacking people but filled with danger and devilment is evident in Christian writing, their own wildspeak as it were, as early as the fourth century. As before, it was described as a place of barbarism, but over time it became sought out as a place in which to reclaim a kind of purity that was understood to reside in the world outside of humanity. It was where monsters against God could be found, but simultaneously offered an escape for people who wished to reject the sinful progress of humans and their urbanism and live alongside animals, untainted. One such seeker was a hermit born in the late eighth century, Saint Guthlac. Having grown tired of the beer-drinking monks of Repton, Derbyshire, Guthlac is said to have travelled east in search of solitude and arrived at the Fens, a place described in accounts of his life as 'hideous', of 'huge bignesse', and 'clouded with moist and dark vapours'.[3] Here he will have encountered ducks, wading birds and eels, marshes and reeds. It would be easy to imagine this place, so like the peaty bogs filled with prehistoric bodies and associated with tales of blinding mists and witches, empty of people, who might have shunned its wet and dangerous land in favour of higher and safer ground. But once there, he asked inhabitants where he might find the most isolated wilderness of all and was duly pointed to a hill that was *so* wild, he was told, no-one could endure it for they would be plagued by spirits in their dreams. Sure enough, Guthlac was visited from the earth and the sky by fire-spitting beings with 'crooked jaws', 'stinking mouths' and 'preposterous feet'. Many readings of Guthlac's life see his eventual victory over the demon visitations as an allegory for human and religious triumph over a wild and demonic place that had hitherto resisted control and domestication by people at the behest of God.

The Fens were not, of course, a place devoid of people even in the eighth century: Saint Guthlac, our friend in search of solitude, asked the inhabitants where he could find the most isolated place he could live, and that place turned out to be a barrow, or burial mound, which is a clear indication of past human interaction even if it had become relatively remote. The barrow being a remnant of the past and the dead was important *because* it was haunted, allowing Guthlac something to rally against. In fact, the Fens were somewhere that had afforded many people food and a living, perhaps as a thatcher, a fisherman or in waterfowling, and had been for a long time. We know the Romans had managed the area to an extent, creating a number of navigable waterways. Although it remained a wet and marshy place, perhaps the most telling parts of Guthlac's entire story are the 'dark vapours' and 'foul' brackish appearance of the land. Far from representing an untouched wilderness, it's more indicative of a damaged ecosystem that had been unsettled by human activity but was slowly being reclaimed as old drainage endeavours began to fail.[4]

The medieval woods

The area I grew up in on the Wiltshire–Hampshire border was surrounded by woods, and I spent a lot of time in two in particular:

Bentley Wood and Hound Wood. They're both thought of locally as remnants of something more ancient. Hound Wood is mentioned in thirteenth-century documents as *hundewude*, so the name is unchanged in 800 years. Over time, chunks of the wood have been removed entirely and replaced by arable crops in a process called 'assarting', and it has been left in an irregular shape, with fields cut out of the edges like a caterpillar-nibbled leaf. Assarting was effectively illegal in somewhere designated as a royal forest, which we'll discuss below, and required permission of the monarch. This was because arable fields are great for providing only one type of food, and cannot compete with a wood that could function both as habitat for game as well as providing food for people, grazing for stock, and timber, coppice or charcoal products. There is potential in this small patch of wood, now just under 400ac. It's carved up and contains compartments of different types of trees – there is some plantation, inevitably, where old tree cover has been replaced by productive but ecologically uninspiring imported pine. As with Białowieża, using LiDAR survey results here allows us to 'see' the archaeology through the trees.

Sure enough, sifting through the available data, Hound Wood is stuffed full of sites, many unrecorded. There are old ditches and boundaries, possibly delineating bits of wood that required protection in the past. Quarry pits abound too – large bowls, tens of metres across, from where chalk was extracted for use in building or liming and fertilising. These are pertinent: they exist both within and without the woodland and across the area there are many of them. Outside of the wood they're much fainter because they've been ploughed out over time. However, they're so big that it is highly unlikely they would have been dug into productive land *while* it was under cultivation. I think, rather than denoting an area that was open at the time they were dug, they were probably dug in what was woodland. If this is the case, the faint pits outside the wood might denote the old extent of woodland that has since shrunk.

The survival of deep pits in the woods doesn't necessarily mean that Hound Wood is thousands of years old, though it is feasible. I believe it would have been contiguous, originally, with nearby Bentley Wood. Bentley has these same pits, but underneath all of it is a whole lot more than some old paths and wood banks. There is an extensive prehistoric or Romano-British field system, which can only mean that the area has not been continuously wooded but has regenerated after the fields were abandoned. The Roman road between Sorviodunum (Salisbury) and Venta Belgarum (Winchester) runs close by, but there's a range of prehistoric activity in the area too.

One measure we can use to try to get to the bottom of issues like this employs what are called Ancient Woodland Indicator (AWI) species. I'm sure many of you are familiar with the oft-repeated claim that bluebells mean a woodland is ancient, but one species is not really enough to deem it so. There is a woodland near me that is renowned for its bluebell display and attracts many visitors with good reason: in spring the understory is powdered with the iridescent periwinkle of these delicate flowers. But I came to know the site for a different reason: it's also in the area of an old landscape park and I was curious to learn more. What I found was that, until 1801, the 'ancient bluebell wood' was actually half a village, with all the buildings and garden plots and people that went with it. At this point the landowner evicted everyone into some newly constructed houses elsewhere in the village, built in 'Swiss Cottage' style so as to fit the aesthetic he was going for. Then he demolished their old homes and promptly extended his landscape park into what became this not-so-ancient wood. The park was known by the Victorians for its 'sylvan beauty' and various owners undertook a lot more besides, damming a chalk stream to create a series of pools and cascades, and constructing one of the largest walled gardens I've ever seen. It was not until the early–mid-twentieth century that the park was effectively abandoned, and the house that was attached to it first partially burned down and was then converted to a prison. The park became the property of the

Ministry of Defence, which didn't manage it in quite the same way, and only then did beech woodland (re)generate on the site. Bluebells joined, but must have been there for less than a century. There are a smattering of beautiful veteran trees like chestnuts ringing the park, and these are clearly anthropogenic. The steep hill behind the park was deliberately wooded to give a sense of the estate being larger than it was, when in reality on top of the downs beyond the narrow band of trees was common land for sheep.

One study on AWIs has thrown up some very interesting results: they really can be indicative of ancient woodland but this may be the case *even if* the area has had significant periods of deforestation, traced by sediment cores. Some are hardier than has been thought and can survive in the absence of woodland cover, particularly if they are provided shade by other factors, like being in a hollow. It has even been shown that some of these plants exist in areas that lost their woodland *in the Iron Age*, over 2,000 years ago. On the other hand, some ancient indicators actually point to other types of ancient habitat entirely, such as grassland, but are also able to grow in woodland. On treeless Skomer Island, off the Pembrokeshire coast in Wales, three times as many indicator species were identified than at Derrycunihy Wood in south-west Ireland, which has been shown to have been continuously wooded for at least 5,000 years, albeit punctuated with burning and other human management. Despite the high number of indicators and the fact that its name is interpreted as 'Wooded Isle', coring shows Skomer to have been alternating between heathland and grassland for at least 2,700 years. Researchers also discovered that some plants are indicators of *recent* woodland: ribwort plantain, the plant of disturbed ground, as well as pignut – a ferny kind of wild celery – and stinging nettle.[56]

The reason all this matters is because an old wood's value isn't just its age. The value is in the particular blend of plants and animals that makes up the whole – it is a community, and with that comes much more complex developmental histories than we will ever understand.

Most of what we think of as old 'wildwood' may have been formed through the interplay between humans, animals and the environment over time, resulting in a mixture of plants that have come to rely on that history. From the Mesolithic, when we just don't know how some of the ecosystems worked at all, to the medieval period, there had been huge change. It is true for even the most 'ancient' landscapes that we think have survived untouched, like Ireland and Northern Ireland. The woods and treeways in Northern Ireland have been used in evocative scenes in popular television shows and the country has enjoyed a tourist boom as a result, but actual old woods were few and far between by as early as the ninth century AD. At this point there is written evidence in which the three woodland 'wildernesses' of Ireland are mentioned in a way that suggests remarkable rarity. Although we still long for that desolate wilderness without people, the evidence suggests we have been struggling to find it for centuries.

Fossilised fields

The majority of the population of Britain throughout this entire period was engaged in agricultural practice, and broad patterns emerged in the way this was organised across the countryside. There were village settlements that clustered around a core, in the lowland south and east, and more disparate hamlets in the uplands, north and west. In places where there were fewer people, ways of working the land were less communal and so in the upland areas the patchwork tended to be made from well-defined and enclosed individual fields. Across the land there was also 'waste', which was any land that was not used for building or cultivation. It tended to be poor-quality, but was an important communal resource that was rich in forage where it existed, even if not many ventured to make a living from it. In the south and east there were 'open' fields made up of strips with an overall boundary held together by grooves and banks made by plough teams going over the same ground each year. There were no fences, but the banks formed by the repeated throwing up of soil by the plough delineated edges. Farmers could have the rights to disparate strips across the large system of fields that served a particular village. As capitalism emerged and people began to collect land as a commodity – the seed for which was, as noted, initially sown by the Romans – they rationalised their fields by joining their strips – an important step towards 'enclosure' (a long process that culminated in the eighteenth century with a frenetic reorganisation of the countryside that transformed it into the grid-like view of green and brown we see from the hilltops today). These new fields were sometimes then hedged or fenced along their boundaries. Both these and the remnants of the ridges and furrows of the plough can be seen today across Britain, although many that remained under plough have been flattened by mechanised equipment. Their boundaries are long and sinuous, snaking into a gentle curve that fossilises the action of the plough team, which had to begin their turn at the end of a strip early, and thus formed not straight lines but reverse S-shapes. The ridges and furrows in what are now grassy fields can be pronounced, and can often be spotted from trains through the countryside, especially in low winter light when frost has formed and dusts the shady side of the bank.

The long ridges and furrows that formed with the plough fundamentally altered what 'naturally' grows now in many of these places. The furrows are always wetter, ideal for particular crops such as oats, with wheat and peas grown in the higher and dryer ridges. Now, they can be dominated by reedy grasses, or by meadowsweet with its clouds of tiny white flowers in irregular clusters. This often also grows as a 'weed' in roadside ditches and old meadows, and smells medicinal because it *is* – rich in salicylic acid and the base for developing aspirin, so named for the plant's old botanical label, *Spiraea ulmaria*.

The extension of agriculture into wet places is sometimes hinted at in field names, and we shouldn't be surprised when these places eventually flood. Examples include Withy Mead and Rush Bed, indicating damp tracts filled with willow and reeds, or really any name containing *marsh*. Some of the issues we face now with frequent flooding might perhaps be helped, in some cases, by looking to the medieval past as a model of how these landscapes were designed to work. Where we now tend to assign one role to a field, even if it's in some kind of rotation system, the diversity of its use would have been far greater in the past, just like forests and heaths and moors that were multifunctional spaces. Commoners often had the right to graze their animals on the stubble of medieval fields after harvest and before it got too wet, when hooves would damage it for the following year. The division of fields and forests into one thing or another – for amenity *or* farming, for grazing *or* people – is a much more recent phenomenon.

There are very rare examples of places that still farm in the medieval way, communally and in a three-field rotation. These places, as at Laxton in Nottinghamshire, are special not just because they preserve a cultural heritage, but because they also show us the abundance of biodiversity that was made possible with old farming regimes. Yes, we had altered the 'natural' landscape, but this allowed new types of diversity to begin to flourish in the habitats that were created. The unmanaged grassy verges of Laxton's banks are so rich in herb and grass species that they are considered the best remaining example of neutral grassland in Nottinghamshire and designated as Sites of Special Scientific Interest (SSSIs) for their variety, including pepper saxifrage, quaking grass, red fescue, lady's smock and adder's-tongue fern.

Part of the way the land worked was by taking advantage of the opportunities that it afforded without excluding all others, though that doesn't mean that there weren't challenges. Farming had expanded, yet again, with a climatic anomaly that brought warmer weather for a period of around three centuries up to the beginning of the fourteenth century, and the population grew. Farming once again moved to high ground that wasn't exactly the most suitable for growing arable crops, such as steep-sided chalk downland scarps or the

Pennines. These appear as 'steps' running along steep hills, and while we now know from OSL dating that some of these are actually prehistoric, others represent the peak of medieval farming expansion before the economic and social downturns that followed, most notably caused by the plague, affected population levels. In the ebb and flow of agricultural activity over the last 6,000 years, these lynchets represent something of a spring tide.

The forests of kings

There is a tiny village in Derbyshire, a county dominated by the Peak District, all gritstone crags, hills and plateaus, broken by deep valleys through limestone faults. Its evocative name is Peak Forest, though if you were imagining a collection of thatched cottages surrounded by thriving greenery I'm sorry to say its name is rather less reflective of its current surroundings. The settlement and parish are quite late creations in the grand scheme of settlement in Britain, seeming to have sprung into existence in the seventeenth century. This was followed by a flurry of activity that has left its mark: nineteenth-century farmsteads, and extensive evidence of lime-burning for fertiliser. It's possible then that this was all carved out of a remnant wood, and so named for its impressive surroundings. But just as Peak Forest isn't surrounded by forest at all today, the name isn't even particularly representative of its past surroundings either, not in the way you might think. This tiny hamlet's name is all that is left of the medieval royal Forest of High Peak.

The word 'forest', in common with many in the English language, derives from Latin – possibly the term *foris*, meaning 'outside'. It doesn't seem to be ancient, though. The term *foresta* first appears in a legal document of 648 under the Frankish Merovingian ruler Sigebert III, which refers to an area that was considered a 'vast solitude' and home to wild beasts. In medieval Britain, shortly after the Conquest, the concept of 'forest' was not applied to physical woods but introduced in Norman law to denote a large area of land under royal jurisdiction and legally removed from common law: royal forests. Some of the most famous of these royal forests are places like the New Forest and Sherwood, which admittedly were largely wooded. Not all had lots of trees, and the Forest of High Peak was largely one of moorland, made up of open areas punctuated with steep wooded valleys. Pollen analysis shows that by the time High Peak was designated as royal forest, it was already denuded of much of the tree cover that had been present in the past. There seems to have been some recovery of trees in the first part of the later medieval period, in the twelfth and thirteenth centuries, which is likely due to the way that these 'forests' were managed to provide both habitat and timber.[7]

The reason that we have come to associate the word 'forest' with woodland is because they were managed for hunted animals that do best with access to woods and trees. It isn't quite clear, however, when the association between *foresta* and hunting became so strong. The Merovingian kings who used *foresta* seem to have been more preoccupied with making it very clear which land belonged to them. Wild animals were an extension of the land, deemed to belong to whoever owned the soil on which they stood.

When William I landed on English shores and enforced Norman rule, he found a land that was already partly managed as an arena to ensure a good hunt. An earlier king, the Danish Cnut, is often reported to have implemented England's first laws making the killing of certain animals an aristocratic right five decades before William entered the scene. This, however, is largely agreed to be based on a twelfth-century forgery, fabricated to prop up the rules that came with the Norman Conquest. It was about more than who was good enough to kill wild things – it was believed that poaching gathered people together when they should be at church and led them astray, plotting to overthrow the rich in peasant-led revolts. But there were certainly already elements of the landscape here that were enhanced with the specific aim of hunting deer: not via the park menageries of the Romans, but via something called *haga*. There is some debate about what form they took – some believe them to have been bounded enclosures for keeping deer, while others see them as openings in boundaries that could be used to drive deer into nets to catch them. There is one named 'the white *haga*' at Faccombe Netherton, first evidenced in the mid-tenth century, that is traceable now as a low bank with flints that would have made it appear white on the surface.[8] It is now thought that this feature was actually an enclosure and therefore one of the earliest deer parks in Britain. We'll return to parks below, but what is most interesting about *haga* features is that their known locations do strongly correlate to places that came to be protected under Norman forest law. So, although the Merovingian notion of forests hadn't originated from a concern with hunting, it was hunting that became a prominent concern after the Conquest, possibly due to the recognition that the woods were already much harmed by agriculture and wouldn't be able to support the animals that were required if they weren't protected. Roughly a third of England itself was designated forest by the early thirteenth century, and both Scotland and Wales had their own forests. It would be a mistake to think of them as peopleless tracts of trees in which the only thing that happened regularly was the royal hunt. Not only did the royal household rarely visit most of them, but hunting wasn't even the main activity that took place on a day-to-day basis.

This stems from the very nature and physicality – or lack thereof – of forests. Forest areas were not enclosed with fences or banks and ditches, and they varied in size from only a few hundred acres to entire counties. Many royal forests were contiguous, one running into the other with nothing to mark the change, and almost all of them belonged to the monarch, with a rare few along the Welsh Marches under the control of local lords. They could encompass everything from heath, moor, wood and wetland to managed farms and villages. People weren't banned from entering them necessarily, but they were heavily restricted in what they could do there.

Forest law protected two things: deer and trees. The most serious offences involved stealing or killing deer, or things which removed or damaged the *vert* – the trees and browse on which the deer depended. People could be punished for allowing their livestock to 'escape' into protected areas, but were also allowed to fence around their crops to protect them from wild animals. Even possessing dogs was regulated, with owners fined for refusing to bring forward their animals for the regularised practice of 'expeditating' them, which involved declawing and sometimes removing the pads of the feet entirely to ensure they remained of no real threat to the deer. Great portions of the land were given rules that could sit alongside or even contradict existing common law and customs. In fact, even if someone owned a patch of land privately in their own right, it and they were subject to the laws of the forest if they were within one.

All of this rendered forest law, as you can imagine, deeply unpopular. A large number of rural communities were affected, losing the freedom to access and forage from the land as they had in the past. It had been entirely normal to gather plants, herbs and animals for food, medicine, building or other products – something that is quite alien to us in contemporary society. We live under a perpetual cloud of confusion about what we are allowed to do in any given place, affecting everything from outdoor swimming to garden management, foraging and footpaths. Most people are never quite sure if they can even cut back a hedge that is invading their garden (you can trim back overhanging branches, but not uproot a tree). You can gather fruit, nuts and plants that are 'growing wild', but whether this would include a self-seeded crab apple tree on the outside of someone's garden hedgerow is debatable. Is it wild enough to be allowed, and if so, is it actually protected and therefore *not* allowed?

Anyway, the forest was a busy place, filled with people who needed to make their living or simply survive on the resources the land – whether moor or wood or field – had to offer. Thus, despite only coming into force in the decades after the Conquest, a century later it was clear that the existence of forest law was detrimental to many. Magna Carta was signed in 1215 and contained specific clauses dealing with forests, later taken out and put into their own Charter of the Forest in 1217. The result was that some areas were 'disafforested' entirely, and in other cases rights were provided for those living within forest areas. Examples of these rights have famously persisted despite the collapse of the forest law system after the Tudor period. In the New Forest, commoners still have rights of pannage (to graze pigs on beech mast or acorns in autumn), pasture (to graze livestock) and estovers (the right to take wood, heather or bracken for fuel). Other rights that have been extinguished but existed within the forest were those of turbary (using peat for fuel), piscary (taking fish) and marl (extracting stone, clay and gravel for building).

Despite these moves to allow residents to live, rather than simply exist, within the royal forests, people could still be mutilated or even killed for their part in 'evil' offences against deer and there were numerous staff tasked with upholding the rules. The most serious of crimes went through the Forest Eyre, which was supposed to be held every other year, but in practice a quarter of a century could pass before the next one. The result was that someone could be 'maimed' for having killed or attacked a deer many years before the court was even held. The Crown had a long reach and an even longer memory.

Beasts of the forest and park

In effect, forest law was an early form of habitat conservation for the animals it protected, a bit like nature reserves are now, but medieval forests didn't resemble the wooded utopia of Snow White or Sleeping Beauty. They were never created for the sake of the animals but for the sake of the rich, and despite boosting the numbers of wild animals for a while, the boon couldn't last forever.

The animals that were protected by law were called beasts of the forest, the chase or the warren. The wild forest animals were those held in the most esteem and included red deer, boar, wolves and foxes. Rights of the warren and chase often went to noble tenants or ecclesiastical houses and were large areas of land where these people could hunt lesser wild animals like the roe deer, badger, partridge, hare and pheasant. There was a clear social scale in terms of human–animal interaction – the best types of hunting and most prestigious animals reserved, and other types regarded as lowly or even criminal. Male red deer, harts, were reserved for the long chase, known as *par force* hunting, that had been popular in Norman lands before being transplanted to Britain. Fallow deer, smaller and spottier, were suited to enclosed areas. It is possible then that places within the landscape were manipulated in different ways in order to suit a variety of hunting methods, a range of ways to make sure the landscape was well stocked.

Rabbits may be considered wild, but they were introduced by the Romans. Even though they are able to breed and spread very quickly, they had not become used to our climate and don't seem to have survived into the early medieval period, requiring reintroduction by the Norman regime. Until the eighteenth century, they required artificial underground 'housing' in the form of mounds that look a lot like Neolithic long barrows – lozenge-shaped heaps of earth in which they could burrow for warmth. In the early years of the Norman regime, rabbits were a rare and prestige animal and may have been kept first on islands, before being found in royal settings. The very earliest Norman examples we have now date to the thirteenth century, both from palaces where they were kept in enclosures for protection from predators – and from people. Confusingly, the enclosures themselves are also called 'warrens', as are the tunnels the rabbits make underground.

Aside from the forests, the most obvious forms of wild-animal management were the parks, the spiritual successors to the Roman game reserves and menageries. They were distinct from the forests though many were made inside areas under that law. There were thirty-seven parks noted in the eleventh-century Domesday survey, although we're not entirely sure of their form.[9] The survey wasn't designed to note absolutely everything that existed in England and Wales; it was more an exercise in understanding what wealth existed in this newly conquered land. Whatever the true number in the eleventh century, within 200 years park creation reached its apex and there were hundreds across Britain.[10]

To many, a park is what we see surrounding the great estates of the countryside, a peaceful area of rolling grass and trees with huge canopies surrounding a sprawling country house, perhaps with a lake, sometimes with rustic ruined grottos and certainly with a herd of deer, either now or in the past. These types of parks are there to look nice, functioning as rather oversized pleasure gardens, but they don't really reflect the reality of medieval deer parks, which had many functions. They were there to provide deer for a landowner, as a supply of venison for use as food and social currency. They also looked nice, but represented a nefarious creeping trend – that what the later medieval lords really wanted was to exclude peasants from their land. Parks demolished any common rights that had existed, taking away the rights of ordinary people to forage firewood or graze their pigs. While forests could still be considered as places of living and working, parks could be devoid of people other than those directly employed in their upkeep. They were also so valuable that the monarch had to grant a licence just to make one, because this meant the land and deer were also removed from the monarch's wealth. They provided vital sources of timber – necessary because there was no longer enough common wood to provide for the needs of timber-hungry estates – as well as livestock grazing and even farmland. Parks were enclosed not with the brick walls we often see alongside country roads nowadays but by enormous, deep ditches, to the outside of which were high banks topped with wooden pales to form close-jointed fences. They were traps that deer could jump into but not out of, with cleverly designed 'leaps' that made it even easier for deer to stray into them. All of this careful management meant herds were kept stocked for hundreds of years – some of them existing to this day.

To manage a healthy herd of deer and keep the ponds stocked, the coppice wood worked and the lawns and paddocks in order, activity within parks was constant and they were quite an investment. But the people outside the park were not passive and poaching was constant, in part because these animals were supposedly wild and people felt keenly their right to take them, theft also acting as a form of protest. These high-fenced and well-tended parcels of land were an ostentatious display of wealth and control that took land away from the masses, and were run for the benefit of people who often didn't even visit the many estates and parks they held.

I have excavated one of these parks, at the medieval palace of Clarendon – extreme with regard to the massive scale of the park and the buildings within. The assemblage there demonstrates the extent to which people would go to keep and use the landscape for managing wild animals. The site was transformed from a simple Norman hunting lodge into one of the most important secular buildings in all of southern England by several hunt-loving kings, particularly Henry I, II and III, who ruled in the twelfth and thirteenth centuries. It is a stunning site, overlooking the spire of Salisbury Cathedral, and set within what was the biggest deer park in the country at one point, deep within a forest that had been renowned for its hunting since at least the early medieval period. Here, there are upstanding remains of the walls of a Great Hall, huge stables, a large wine cellar, apartments for the king and queen, and even the remains of garden terraces, all cared for by a community group who ensure the palace is not lost once again to the scrub that claimed it for hundreds of years after it was abandoned. Still, so much about the place is unknown that in 2019 two trenches were excavated.

Alongside the evidence for previously unidentified buildings, the most interesting stories emerged from the assemblage of animal bones. In recovering the material, and carefully sieving samples of the sediment in each trench, we were able to uncover evidence for the kinds of animal life that were present. As expected, the proportion of wild-animal bones in the overall assemblage was higher than would be expected in a lower-status site – we already know that the wild was of high status at this point. But the sheer scale of the assemblage was surprising, even to the animal bone specialists who analysed the collection. Whereas an elite site might be expected to have around 15 per cent of the animal bone deriving from wild sources, at Clarendon the rate was nearer 40 per cent,[11] reflecting just how important the site was as a retreat for the monarchy but also how symbolic wild bodies were. Most of it was deer, as befits the palace's placement in a huge park, but some unusual animals were also present, such as crane, a high-status bird that would have been incredibly rare at this point in history, as well as evidence of the consumption of flat fish, such as turbot, that must have been brought from the south coast. In one set of related layers there was even evidence for what may be the earliest medieval rabbit in Britain, and one of the last wildcats in England, thought to have been long extinct. The presence of the wildcat, alongside fox and badger remains, hints that the rabbits here may have been farmed. Predators do not usually form food waste, so their bones represent efforts to eradicate them.

The ecological impact

All of this wild management had an impact on and within the landscape. We chipped away at forests but also protected tracts of them, heavily managing who had access to what. Intensification of farming crops and, later, the growth of the sheep-farming industry led to significant ecological impacts in Britain. Wild animals do not occur with great frequency in the archaeological record for the early medieval period. In some cases, parts of animals, like bear and beaver, accompany burials or cremations and may have been used as amulets and seen as holding some kind of power. In places with wetland there does look to have been more 'harvesting' of wild resources, such as fish and crane in addition to beavers. Regional use of these areas makes sense, particularly if previous Roman drainage was failing and cultivation was becoming more difficult. They were abundant sources of forage for those who could, and needed to, access them. There does not seem to be much evidence for *over*-exploitation in the early period, or either introductions or extinctions, until after the Conquest. Many of the animals we've lost in the wild had already gone before the medieval period, but some of the larger ones are thought to have survived prehistoric and Roman management and remained. These – the bear, beaver, boar and lynx – gradually disappeared over the medieval period, and all had gone before Henry VIII took the throne. Others, like different types of deer, suffered population decline and the decimation of their range.

Bears were probably extinct before the Conquest, likely due to a combination of habitat loss, deliberate extermination to protect livestock, and hunting for sport and status, but the picture is complicated by evidence left over from city-based bloodsports. There is very little archaeological evidence for bears living in Britain in later prehistory or the Roman period, with just one bone recovered from excavations at the amphitheatre in London. It has been suggested that bears were either present in extremely low numbers in the wild during these periods or already extinct. A greater number of bones have been recovered for the early medieval period, but these are not from whole skeletons and suggest a trade in parts, such as skins or bones as amulets. After the Conquest there is a shift to evidence reflecting the presence of live animals, which were likely imported to fulfil a role in entertainment – bear- and bull-baiting were popular pastimes for some in the medieval period. By the fifteenth century, there was a royal keeper of bears, but they are rare in tapestry, paintings and other art.[12] During the Tudor and Elizabethan periods bear-baiting was staged in its own special arenas – a little like the Roman Colosseum, though smaller – known as bear gardens, which often doubled as theatres for plays, so that on one day an audience could see Shakespeare and on another watch a bear being set upon by large dogs.

The beaver – hunted for its fur and castoreum, and likely to be persecuted if interfering with managed wetlands – may have clung on for a lot longer than the bear, possibly only disappearing from remaining wetland habitats in the fourteenth or fifteenth century. Written records indicate their presence in the eighteenth century, but these may not be accurate. Often, sources borrowed heavily from each other and perpetuated assertions based on little or no evidence. The watery places were also undergoing changes, and fishing was transformed from the early to late medieval period, too. The archaeology dating to earlier than the eleventh century is made up of freshwater species such as pike and eel. But the later assemblages show a massive increase in marine species like cod and herring, believed to represent a moment in time that marks the start of overfishing that has left us, 1,000 years on, with barely any left.[13] It is also indicative of growing long-distance trade networks and the commodification of fish stocks. Cured fish became a bulk product, traded and consumed far from the sea. Fish, particularly salted marine species, were consumed during Lent – but it's a myth that in the seventh century Pope Gregory the Great decided that baby rabbits were fish and could therefore be consumed. The ruling has often been linked to the spread of rabbit farming, but it doesn't seem to have any basis in truth. The Normans constructed fishponds on their estates, but they were largely to convey status and the fish kept in them were eaten only on special occasions. Carp, although nowadays the most valuable sport fish and not considered edible by most, are not indigenous and were introduced to these ponds in the fourteenth century as food. We also find the first evidence for fishmongers, who often kept their fish in 'stews' – ponds where fish were livestock, like giant versions of the tanks you see at modern fish markets with live lobsters or crabs on display. In medieval London, stews were situated along the Thames for easy access and distribution. In the landscapes surrounding manors and castles, fishponds were managed and the right to fish in them and along riverbanks was protected, just like with parks and forests. Unlike hunting, however, fishing did not become a sport until the eighteenth century.

The forests and parks inevitably had an impact on animal populations. Although after forest law appeared we see a rise in the number of wild remains in archaeological assemblages, it wasn't to last. Initially, habitat protection and close management of park populations meant that there were more deer, and numbers were bolstered by the introduction of fallow deer in the twelfth century. The fallow deer quickly became the favoured park deer, so well suited they probably partly drove the creation of the parks in the first place. They did well, but others did not. By the fifteenth century the smaller roe deer was a rare sight everywhere other than the north of England and Scotland.

Other animals of the forest fared no better. Wild boar, an icon of the medieval forest on the continent, seemingly disappeared in the thirteenth century. In fact, it's been suggested that the reason the boar was not as revered in medieval Britain as elsewhere in Europe could be because they were already so rare that the opportunity to hunt them evaded even royalty. A continuing depletion of woodland even in protected forest areas played its part, but boar are also more difficult to enclose and control than deer and it is highly likely local populations hybridised with domestic swine. In effect, they were diluted out of existence. The last of our predators, the lynx, also suffered. These were reliant on woodland-grazing ungulates but don't seem to have benefitted from the short-lived uptick in deer populations as they were already all but gone by the Conquest. They are archaeologically extremely rare even in the early medieval period – only a single specimen has been identified, from fifth–sixth-century Yorkshire – and a lack of cultural and artistic depictions suggests they were also unknown later on.[14]

The birth of legend

Where wilderness had its own hold on the medieval mind, arguably the medieval countryside itself has its own hold on ours. So much is tangible in the buildings, castles, fields and churches that we cannot help but be a bit swept up by it all. These stories were born in the medieval period because of the way in which people at the time were coming to view them. Wild animals were simultaneously claimed as the property of the rich, but declared wild and therefore unowned; they could be protected for hunting but not considered edible; and labourers had to produce grain for their landlords even though the land might be taken away and turned into a park. Multiple identities thus befell the space outside the village or town. Nature was God's pure and unsullied space, but also the monarch's to denude at will. Against this kaleidoscope of landscape ideals, folkloric stories began to emerge to make sense of what was going on. It is common to encounter placemaking myths even now, attached to roadside crosses or topographic features – witches said to live in caves and ghosts of the hanged at old crossroads. Many of these are localised, but share common traits.

Nestled in the base of a valley not far from the World Heritage Site surrounding Stonehenge and Avebury is a small monument known as the Bloodstone. It's a smooth chunk of sarsen, the type of stone used to form the famous trilithons. Sarsens occur naturally in this part of the world, the remains of a crust that formed over chalk when groundwater carried silica through the layers and cemented them together. The crust fractured and the boulders drifted around with periglacial activity, or were sometimes moved by people. Sarsens have had many names, such as 'greywethers' for their resemblance to a flock of sheep from afar, and 'breeding stone' because they often appear clustered together. The word 'sarsen' itself is thought to derive from Saracen, which was more commonly used in the medieval period to refer to Muslims. In the context of the stone it is suggested that the word was applied because of sarsen's 'otherness' in relation to the underlying geography.[15] In medieval stories Saracen characters were often either quick and willing converts to Christianity or, particularly in later work, described as monstrous villains. Stories came to attach themselves to these stones that seemed to be out of place.

The Bloodstone is of sarsen but much smaller than the giant Stonehenge examples which are covered in lichen, with hollows that harbour nesting birds. This one has a rusty patina, a red-hued rind over the surface that has been caused by its iron content, and lies within the ditch of a Bronze Age barrow. There is a line of barrows here that leads to a boundary that has possibly persisted since prehistory, using the head of a spring as one of its naturally 'fixed' points. The name Bloodstone refers to its colour, which according to legend is staining from the blood of Vikings who were beheaded on it following King Alfred's decisive victory at the ninth-century Battle of Edington. Fuel was added to this fantasy in the 1950s when a burial was found near the spring below: a skeleton placed face-down, with its hands positioned so they look to have been tied behind the victim's back. Other skeletal remains followed; however, a medieval date has eluded scientific analysis. All have been shown to be either Roman or several centuries later than the famous Saxon–Viking battle. Nonetheless, the power of the Bloodstone and its myths lie in the stories we tell about the landscape around us. The monuments of earlier cultures were very visible in the medieval period, both recognisable, in the case of Roman remains, and also less understood, such as the Bloodstone and burial mounds. These appear in early medieval charters that define the boundaries of settlements, monasteries and abbeys, but also in some of the earliest stories that were written down in this part of the world.

The story of *Beowulf* is set in sixth-century Denmark and Sweden, though may not have originated until the eighth century, and the only surviving written form is from around 1000. In it, the hero Beowulf successfully defeats a monster, Grendel, and its mother. Years later, when he has become king he must fight a dragon. The terrifying dragon lives in a barrow, referred to as a 'stone barrow' and an 'earth cave', where few dare to try and enter. There is also treasure there, suggesting that the people telling these stories were familiar with the precious grave goods that could be held inside such places.[16] In Irish mythology, barrows and cairns were often seen as part of the world of the *fae* (fairies), an invisible but significant presence in the countryside. In effect, they had become part of something outside the human realm, taken in by a wild magic. These fairies were *sidh* who inhabited an underworld of rich palaces, taking the form of leprechauns, banshees or shapeshifting goblins known as *púca*. All sorts of features including springs, lone trees and ringforts were associated with the *sidh*, and tradition warned that those who interfered with them would meet their end.[17] Whether or not people believe in the fairies, ringforts are still spoken of as places to be avoided, and many remain unexcavated, perhaps because they are locally protected from prying archaeologists.

Supernatural forces and monsters were thought to dwell in the sea as well as on land. Like bogs and their tendency to fool, the sea is seldom to be taken at face value and requires specialist knowledge to navigate safely. Unlike bogs, it moves, forming waves and currents, sometimes invisible under the surface. Medieval bestiaries – books that described real and not-so-real animals and their religious meaning – are filled with sea monsters, though not all are dangerous. Sea serpents are depicted frequently on ocean maps, and there are many hybrid-type animals that take the form of a fish mixed with something from land. The 'sea-ass', the sea-dwelling counterpart of the donkey, is shown with a long, scaly body and the head of a donkey, and is said to be able to cure people of 'frenzy'. The creatures that lived in these dangerous waters were poorly understood, truth be told. Small whales were certainly caught, but larger ones were scavenged if they washed up on shore. Large marine creatures like giant squid, normally only found in the deep sea, which was inaccessible to people until much more recently, may have only been glimpsed as carcasses that had floated to the surface. Not only did people have to consider the state of the weather, of what may come at them from the sky, but they had no real idea of what was beneath them at any one time, or whether the waves they could see on the horizon were actually a surfacing pod of whales that might capsize their boat with a flick of their tails. Stories that came back with sailors who either fished or, in the earlier period particularly, voyaged across the North Sea must have been as fantastic as they were unverifiable. Here was a zone that was *entirely* unknown to most of the population of Britain, who could only wonder what lay beyond the horizon.

The *folk* in folklore

More tangible in everyday life were tales of specific people and beings that came to represent what it meant to inhabit or be of the wild places. One medieval figure of folklore who is often held up as an exemplar of how people feared the forest is Herne the Hunter, leading his so-called Wild Hunt. He is often depicted as a ghost or figure riding a large horse, with antlers topping his head, his hunt of spectral figures forming a sort of army of the night. Much of the myth is a modern retelling, but it is a powerful image and bits of it really do seem to have their roots in medieval beliefs. The Wild Hunt first appeared as a term in the work of Jacob Grimm (one of the Brothers Grimm), in a book on German folklore. In this, Grimm suggested that Herne the Hunter was an English version of the figure that led the hunt. That said, Herne didn't appear anywhere until Shakespeare included the character in the sixteenth century in *The Merry Wives of Windsor*, in which he is the ghost of a gamekeeper who haunts and terrorises the oak forest of Windsor. Ronald Hutton has suggested that the concept of the Wild Hunt may well be based in medieval and early medieval beliefs about the danger of the night, of the supernatural things that stalked it, but was effectively a conflation of these into an invention by Grimm. These traditions included a fear of nocturnal flights by female spirits, which eventually became the idea of the Witches' Sabbath, as well as fears of phantom hounds that stalked the night.[18] That nocturnal animals like cats and bats are often associated with witches makes sense in this context.

Much folklore was therefore concerned with places that could be explained by their relationship to otherworldly figures, or that were uncomfortable to be in, like underground tombs filled with unknown dead, or the night that was filled with threats of its own. These painted what was outside the settlement, park or town as something to be suspicious of. Other stories, though, were more directly concerned with wild spaces as places of struggle between those who desired to live off the land and those who would seek to own and control it, in particular the tales of Robin Hood. Whether or not Robin Hood was a real person, or even several people, has been debated for centuries, but the basic theme of the stories will be familiar to most. Hood is characterised as an outlaw, living out in the forest with a motley band of others who are often said to have robbed from the rich to give to the poor. The chief antagonist of modern retellings is the Sheriff of Nottingham, whose main aim is to capture and execute Robin Hood, all while Hood outwits him and makes off with his lover Maid Marian. The stories have morphed over time but the earliest literary recordings date to the fourteenth century, and are said to have taken place in the twelfth century. In these, Robin Hood is joined by variations of familiar characters such as Little John and Will Scarlet, and they seem to take place in either Sherwood Forest or Yorkshire. Robin was not a member of the aristocracy in the early days but was a yeoman who robbed from those whom he thought were undeserving of wealth, rather than simply anyone who was rich. The early stories were violent and didn't always paint Robin as a hero – he could be a trickster, mischievous, and frequently killed people. Maid Marian didn't enter the picture until the seventeenth century, when a fashion for chivalric tales rendered Hood a freedom fighter who romantically fought against the Normans.[19]

There are documents that point to potentially real sources for the stories of Hood and his gang. One is a Yorkshireman, William LeFevre, who appears in thirteenth-century sources with the alias William Robehod. Another, Robert Hood, had stolen goods seized from him by the Sheriff of Yorkshire in the thirteenth century. It is also possible that the stories could refer to those who had been labelled as outlaws, using the name Robin Hood as a kind of byword. This scenario would indicate that the *figure* of Robin Hood, wherever he came from, existed in medieval popular imagination to the extent that forest-dwelling criminals could be labelled as such. The real significance of the stories is not so much whether there was a band of men going around stealing from the rich, but rather that they represent the struggle against forest law. It is notable that he is considered an outlaw, when the term *forestis* means literally this. It made the forest, and any who managed to live in and off it, 'other'. The association between the forest as a wild place and criminals goes back much earlier than the tales of Robin Hood, and so the stories themselves may be drawn from earlier traditions. In Old English the word *wearg* was used to refer to both wolves and outlaws, and the figure of the wolf seems to have been used as a metaphor for criminality from the ninth century.[20] Just as Robin Hood rooted himself in popular imagination, the figure of the wolf and what it represented seems to have become more important than the reality. The wolf as outlaw is specific to Old English, which has its roots in Anglo-Saxon society – a culture that largely accompanied people coming into post-Roman Britain from areas where there certainly *were* wolves, ones that stole livestock. It is easy to see why this theft would link wolves and criminals – both inhabited areas outside the social norm.

Transformed

The extent to which there was much wilderness left in medieval Britain is difficult to ascertain, but by the end of the sixteenth century it had come to wear many different faces, some real, many imaginary. Nonetheless, what was left in the wild required control. Some wild animals had to be preserved to provide hunting quarry and stock for the aristocracy and monarchy, which meant persecuting other animals and people that could do them harm. It also required the extensive management of 'wild' habitats. At the same time, farming and timber production required their own forms of protection against animals that could damage them, and so across the land had sprung up myriad spaces with sometimes overlapping as well as competing goals. There were some people who effectively squatted on what was considered waste land, building cottages on the heaths and attempting to make a living from a few acres. What came next set in motion the process that finally carved the rest of the landscape up into parcels, each with their own codes of ownership, rights and perceptions. The world was also about to open up, bringing with it a whole new way of being wild.

5

DISSECTION

(EARLY MODERN, 1540–1800)

A new dawn

The story of how we have shaped the wild arguably started with the wolf, thousands of years ago, when the first of them came close enough to steal scraps of food and the first of us began to overcome our fear enough to build a relationship with them. To a certain extent the story ended, in Britain at least, with the realisation that the last of the wolves was gone. The lessening of fear that drove the creation of the domestic dog all those years ago had driven the last of the old wild to extinction. Wolves had become lapdogs, vast forests had become hunting parks, moors or wheat fields. The enclosure that began in the medieval period shifted in the early modern period to become a principal force in the countryside, gridding out the land into something that much more resembled what we see around us today. New trade networks saw all sorts of new plants and animals find their way to Britain either as imports or as stowaways, with some of them naturalising into our landscape. The native wilds we had lost now became as exotic as the animals and plants we imported as spectacles. The idealisation of Robin Hood that appeared in the sixteenth and seventeenth centuries memorialised the idea that in the centuries prior the wilderness could be roamed not just by wolves and wild boar but also by gangs of outlaws. The fear that we had experienced was a fear of the elements of our world that sat outside of the influence of law and order and that which we did not understand.

The early modern period marks a turning point in the way in which we thought, and still think, about what is wild. We have inherited a reverence for what we consider to be the old wild that sometimes obscures the fact that wild things didn't entirely disappear, they simply *changed*. Something new emerged in place of what was lost. To understand that, we need to explore the way in which we began to relate to what we considered to be dangerous, wild or magical, and the parallel developments in how the world was now becoming categorised and understood. In essence, what was really there and what we *thought* was there diverged. As we recognised loss we began to resent what was there in its place, even if it had introduced remarkable new ways of being wild. We're first going to cover what happened to our thinking about the animals that were left here and what we did to refine them into the breeds we have now, as well as the exotic ones we began to look towards for sources of fear and amusement. New animals required, once again, new landscapes, but this was also accompanied by a yearning for something not quite so obviously human. Artificial wildernesses were assembled on great estates from a variety of old and new plant species. They were to become what we understood the wild to be – safe, beautiful and free from others.

In archaeological terms, there is a distinct shift in the types of evidence that are available from this time. Although it is more commonly recognised as just as archaeological as any other period, historically it was unusual for archaeologists to pay much attention to things that came after the medieval period. It was left to the realm of historians, thought to have so much written evidence that archaeological consideration of surviving material culture wasn't deemed necessary. Everything in this period proliferated, things like mass-produced books, paintings, more reliable maps, the careful recording of accounts and letters that made their way into family papers and eventually into archives. There are now plenty of people who do study the material culture of later periods, but archaeologists also recognised more readily that much can be understood from what remains above ground, in the organisation of landscapes and buildings, plant remains and unusual lumps and bumps. Marrying the history and archaeology together is when things really come into focus, with one able to support or sometimes disagree with the other. Not everything leaves a written trace, nor does everything that is written actually ring true in the archaeology. Many sites I have worked on have never been touched with a trowel or a mattock; we simply mapped and measured, comparing old documents and written testimony with what we found on the ground. This is just as much archaeology as unearthing an arrowhead from a barrow – all of it tells us about the human past.

The last wolf

A few years ago, I sprung a surprise trip on my children. It was the very last day that the grounds of Wilton House would be open before shutting for the season. Aside from the beautifully landscaped grounds that have graced many a Georgian TV drama, it has an adventure playground complete with boat swings, trampolines, treetop walks and perilously high slides. It is the stuff of dreams when you're eight years old. Naturally, I sold this to my children as a treat, but I'd come across something in my photos of archive documents the day before that had grabbed my interest. Having followed it down the research rabbit hole, I had emerged a while later with a lead that was to become my secret agenda for the visit. By that afternoon we were scrambling through a forest of twisted rhododendron trunks in pursuit of a small pet cemetery I had a vague recollection of once finding during my own childhood trips to the estate.

I was in pursuit of a very specific animal, a wolf that for twelve years after 1770 had roamed the grounds of Wilton House, the estate of the Earl of Pembroke and seat of the Herbert family since the dissolution of the monasteries in the sixteenth century. She appears in the household accounts as having been provided with her own keeper, Nathaniel Townsend, who would eventually spend over forty years in the service of the family. Her name was Lupa and at the grand age of twelve she died and was buried with a headstone and epitaph. I reasoned that there was a good chance she was buried in the pet cemetery that I remembered finding as a child. Alas, as much as we rubbed the dirt from the inscriptions, they were either too degraded to read or commemorated animals who had died much more recently. There are no other known records of Lupa held in the private archive at the house, but we do know something of her story. The fact that she was buried with a headstone and epitaph indicates she was possibly more of a pet than a hunting hound, and she was certainly kept separate from the kennelled hunting dogs. She had litters of pups in at least 1773 and 1774, so at some point there were quite a few of these beasts around the grounds. But the puppies were not wolves, because neither, strictly speaking, was Lupa. She was, in fact, at most one-quarter wolf, the product of a mating between a half-wolf and a dog. How did she end up on the Wilton estate, and why was she repeatedly referred to as a wolf in accounts if she was more dog than wolf? To answer this question requires us to travel to London to investigate the man responsible for her birth, the state of natural science, and the understanding of wild things at the time.

John Hunter was a surgeon in London, *the* Hunter of the Hunterian, the museum of the Royal College of Surgeons that houses his collection of biological specimens. Hunter was an avid scientist of not just human biology but also that of other animals. One of his quests was to prove that domestic dogs, wolves and jackals were all the same species. He sought to prove this by mating dogs and wolves together. It was not always a smooth process, with some dogs being mauled and others receiving no interest at all from the wolf. Some of the offspring, including Lupa, were sent out among the nobility: one went, for example, to James Hamilton, Earl of Clanbrassil and MP for Helston, Cornwall, between 1768 and 1774. The wolf–dog hybrids were said to retain something of the wild wolfish characteristics one might expect, and in the case of one of Hunter's own wolf hybrids, a Pomeranian cross, this behaviour led to the animal being stoned to death in the street because people feared it was a dog with rabies. They did not fear that it was a wolf, because they had never seen a real one and would simply never have expected to encounter one. The animal's *wild* behaviour, coming from its non-domesticated genes, was so unrecognisable as wild that it was seen as caused by disease.

We have seen that there is little actual evidence for wolves persisting in Britain in the Roman period and beyond. There were certainly no wolves left in the wild of Britain by the eighteenth century, although the extinction date of this last major non-human predator on our shores is steeped in legend. The last reference to a wolf in Wales appears in the twelfth century, but there were those who claimed wolves lingered in Scotland until 1680 or perhaps even later.

The so-called last wolf in Scotland was claimed to have eventually ended up as a stuffed exhibit in a natural history collection amassed by Ashton Lever. This collection was staggering in size, comprising hundreds of thousands of specimens from around the world. The collection was catalogued when it was sold in 1806 and included both the wolf that was later claimed to be 'the last in Scotland' and another, a wolf–dog hybrid. Lever's collection had been on display in Leicester Square, the very same place where Hunter had opened his teaching museum in 1782: might this hybrid have been a cousin or even sibling of Lupa? Hunter and Lever moved in the same circles: it was a short couple of steps from animal dealer to collector to natural historian, and both Lever and Hunter appear in papers held by the Linnean archives. Hunter's work is referenced in several surviving letters to Carl Linnaeus, with one also suggesting Linnaeus's son visited Hunter's own premises in London. Linnaeus was an eighteenth-century Swedish naturalist, the ultimate cataloguer (as one of the men responsible for the two-part naming system of biological classification), whose work, in part, helps to explain the rise of natural history collections such as Lever's during this period. Linnaeus's naming system gave us a framework for understanding the differences between animals and plants and offered standardisation. This standardisation was great for generally advancing the science of biology and natural history, but it did see a move away from a world in which there was a different way of understanding nature and its inhabitants. The British landscape historian Tom Williamson has pointed out that many wildflowers had localised names across Britain, forming a cultural heritage of stories and uses that was lost when things were brought into line.[1] Some of these names and stories are irrecoverable: they were simply unrecorded and are now completely forgotten. A whole wild world, lost.

Perhaps it doesn't matter when the last wolf disappeared in Britain, but it is the dawning realisation of its loss that is pivotal, the thing that underscored the end of an unpredictable wilderness where humans could be bested. The fact there are so many claims to have seen or killed the 'last' wolf points to its almost mythical status at the time. Lupa was seen as a wolf because she was known to have wild blood and was unlike anything that had been encountered in rural Britain for a long time. In Linnaeus's time the wolf was known as *Canis lupus* and the dog as *Canis familiaris,* and clearly having any real wolf relation was powerful enough to deem Lupa, in some people's eyes at least, *lupus* rather than *familiaris*, regardless of the genetic balance. She was a curiosity, an exotic. Lupa's wolf grandparent made her one of the wildest things that people had ever seen, in a time when the wild was considered, more than ever, to be beyond the human realm. To have wild blood was therefore to be imbued with something *other.*

Werewolves

Even as the world marched towards standardisation and classification there remained a culture of folklore and superstition that pervaded right into Linnaeus's time. While the understanding of animals was changing in learned circles, there were still deeply entrenched beliefs across northern Europe that had perhaps less rational roots and that resisted the new order. The sixteenth and seventeenth centuries are well known for seeing a peak in the persecution of people considered to be witches: King James VI of Scotland, later James I of England, even wrote an entire book on dark magic, the Devil and all manner of beings that could possess a person. But during the same period, we also see a surge in the number of mentions of werewolves, especially across France and Germany. This seems to have been particularly driven by the publication of a German pamphlet which has since disappeared, although two copies survive of an English translation dated to 1590. The pamphlet details the life, crimes and execution of Peter Stumpf in Germany in 1589. Stumpf had been accused of being a werewolf, of witchcraft, murder, cannibalism and incest. He was said to have confessed to a twenty-five-year reign of terror in which he murdered children, adults, goats, sheep and cattle, sometimes in villages and towns but other times in the 'woods or thickets adjoining'. So ferocious and depraved were the attacks that he was sentenced to be stretched on a wheel and defleshed with red-hot pincers, his arms and legs to be broken with wooden hatchets, then beheaded and his body burned to ashes. His head was mounted on a stake above a wooden carving of a wolf as a 'reminder' to others.[2]

Perhaps Peter Stumpf was simply a cannibalistic serial killer, perhaps there was a brave rogue wolf that had developed a taste for human flesh, or perhaps both were true. That the wild was seen as something that could infiltrate the human world and cause such disruption is the most interesting part of the story. Werewolf legends often emerged in places that had suitable habitats for wolves; however, in places such as Britain, where wolves had seemingly become extinct much earlier, anxiety about a legendary wild and the threat it could pose to life and livelihood was seemingly greater than ever. Peter Stumpf's story clearly resonated in Britain for the pamphlet to have been translated and printed, and wolf attacks in France were reported reasonably frequently in British newspapers right into the eighteenth century, especially when the wild was able to infect its victims. On 27 April 1754, the *Whitehall Evening Post* reported an attack in Brittany, France. A young teenage girl and a child had been walking to Mass and had been set upon by a wolf, who seized the child. The older girl ran to seek help, returning to find the child almost torn to pieces. Most of the crowd who returned with the girl were too scared of its ferocity to attack the wolf, except for one man who, so the story goes, hit the wolf on the head until it let go of the child. He held the wolf down while the child was carried away to safety, though the child would die only an hour later. Both the man and older girl had been bitten by the wolf during this episode and although their wounds were not serious, the girl 'died mad' only a week later and the man after a month. During the last twenty-four hours of his life the man was 'raving mad', hallucinated the wolf at the end of his bed and was afraid of water. They had died of rabies. It was as though the disease had infected its victims with the wildness of the host, in its most ferocious and undomesticated state. Such was the fear of rabies across Europe that it contributed to dog taxes, designed to limit the numbers of dogs. England had its own dog tax, though it went hand in hand with other laws designed to limit access, as always, to hunting. Rabies and wolves were both problems far more prevalent on mainland Europe, but held a fascination for the public in Britain.

The wild out in our own landscape was no longer a cause of anxiety in our daily lives, and perhaps this is why our understanding of the real wild became fractured. On the one hand, we had reinvented it in our minds as a peaceful, bountiful utopia, ostensibly unsullied by humans. We had developed an understanding of animals and plants based not on how they were used and related to our lives, but on their anatomy, value and place in a vast network of commodities. On the other hand, the *true* wilderness, as we thought of it, had ceased to be something we ever encountered. We may have quelled the old wild of Britain, but we continued to fear certain elements of it because it represented a threat to the order of things, where we were seen to sit, naturally, on top. When we encountered something that was perceived to be out of place, the reaction was swift – the stoning of a wolf–dog hybrid, for instance – but while it was in its place, we marvelled at it. The wild of elsewhere became something to be bested, and people travelled further afield in search of the thrill of the chase. As we became more adept at marine expeditions, the sea provided an opportunity to obtain and erect huge arches of whale mandibles as signs of dominion over the oceans, and it is no coincidence that taxidermy also became popular during this period. These both served to introduce people to the wild things of far and away, and cement the status of the owner as someone who had access to and control over them. What we thought was wild was no longer something we understood as surrounding us, and had become exotic, fantastical, coveted and marketable.

Breed

It was not until the eighteenth century that breeds, as we know them today, were really created. Before this, there were *types* of domesticated animals perhaps, but the groups were fairly amorphous and changed over time. It was only after a litter of puppies was born that each one was then declared to be of a particular type, those with decent scenting ability assigned to the category of bloodhound rather than to guarding, for instance. Some small dogs were used to turn meat spits in towns, or for rat-catching. It's also at this time that pet-keeping began to become more commonplace, with animals kept for comfort and fashion. Modern breeding methods ensured traceable lineage and the propagation of particular traits of both utility animals and pets, even before a solid understanding of genetic inheritance had been developed.

Breeding methods were partly based on the experience of inventing the thoroughbred breed of horse: all modern thoroughbreds can trace their lineages back to one of three foundation sires imported from the Near East. The farmer Robert Bakewell in Leicestershire also selectively bred cattle to maximise desirable traits and improve the yield of meat or milk. It was his work that led to the average size of cattle doubling from the beginning to the end of the eighteenth century.[3] These animals became projectors of status: the horse for its speed and ability to win a race, the dog for its nose and ability to hunt by scent at the expense of sight, the cow for its meat and portrayal of productivity. This may have been a time before the Kennel Club and Mendelian genetics, but people did understand how they could influence the biology of other beings. So much about these animals had become wholly dependent on the human world.

Exotic wilds

The concept of what was wild was undergoing a shift. While the wild may have been something that we worked with and within, fought and tamed in the distant past, we now had no major predators, scores of introduced species, and dwindling open spaces that were to all but disappear by the nineteenth century.

Animals were a more frequent sight on city streets in the past than they are now: they were bought and sold, worked or scavenged. But from the beginning to the end of this period the types of animals visible to the public changed. Europe colonised other lands from the sixteenth century, a time of exploring the wider world and taking from it what was most profitable or alluring. Animals that had the potential to reveal to European scientists something about their biological mechanisms were brought back for study. Botanical collections were created through the theft of both plants and the indigenous knowledge that went with them. The oldest surviving herbarium, a collection of dried plants for study, dates to 1532 and is still housed in Pisa, Italy. Eventually, though, methods were developed to bring plants back alive and protect them from the salty spray of the sea on long voyages back home. Glasshouses were added to estates in order to grow exotics such as the pineapple, which famously became a craze and synonymous with the very best of the best. Cabinets of curiosity were developed as sort of micro-museums of natural and cultural artefacts from around the world. These cabinets could be collections spanning whole rooms; they were named *Wunderkammer* in German, literally meaning 'room of wonder'. They were intended to represent the sum of human knowledge and included a range of natural and cultural artefacts and eventually gave way to, and formed the foundation of, modern museums. The first public museum was the Ashmolean, Oxford, which opened in 1683.

It is well documented that the legacy of these early and later collecting episodes reverberates today as one of violence against other people, landscapes and cultures. The sugar plantations of the Caribbean were catastrophic for indigenous biodiversity. Barbados was completely deforested within twenty years, taking with it all the species of plants and animals that had relied on the trees. It is impossible to know how many species were lost, and it is now estimated that only 11 per cent of the plants growing 'wild' on the islands are native.[4] On the other end of the scale, a recent study found that after European colonisation of the Americas, so much agricultural land was abandoned – as local communities who were farming it were displaced, murdered or killed by epidemics of new diseases brought by the colonisers – and reclaimed by forest that it caused a global cooling event of 0.15°C in the late sixteenth–early seventeenth century.[5] It is difficult to comprehend the sheer scale of land abandonment for this to have happened.

There were other unintended consequences to this new zeal for collecting and curating, at home as well as abroad. The brown rat, perhaps the most hated pest in Britain, arrived on our shores by 1728. Originally named the Norway rat under the mistaken belief that it had Scandinavian origins, it had in reality spread through Europe from Asia and only reached Norway several decades later. Elsewhere it was generally confined to port cities, pointing to its transportation of choice aboard sea vessels that were locked into vast networks of trade across the globe. In Britain it spread and quickly overran the incumbent population of imported rodents, the black rat, which had arrived with the Romans some 1,500 years previously and had reappeared to spread the plague in the medieval period. The brown rat was the last of three species to become globally distributed, after the black rat and house mouse. Although it didn't cause another human plague, the brown rat quickly made its presence known in other ways. On islands like Lundy, in the Bristol Channel, the brown rat all but wiped out incumbent populations of sea birds, which have only begun to recover in recent decades with the extirpation of the rodent population.

Other species were obtained through specific dealers in exotics who set up shop in order to provide people like Hunter with animals to dissect and learn about, but also to supply menageries or travelling zoos, or estate owners with something 'wild' to put in their landscape parks. They served the zeal for bear-dancing and bear-baiting that was mentioned above, and sourced other interesting animals that had been rarely seen otherwise. In the 1660s Frances Teresa Stuart, Countess of Richmond, obtained an African grey parrot as a pet. Upon her death in 1702 her wax effigy was displayed in Westminster Abbey accompanied by the parrot, now stuffed. It is thought to be the oldest surviving example of bird taxidermy in the world. In all of these cases the animals satisfied people's desire to see wild things, kept at a safe distance or otherwise under some kind of control, and they would pay to do so.

One of the most famous places to see exotic animals in London was the Tower of London. A menagerie of different beasts, from monkeys to lions to polar bears, had been kept there since the thirteenth century, with varying degrees of success. Some of the animals couldn't adapt to our climate and others perished because the keepers simply didn't have the knowledge of how to look after them. Lions, however, were a near-constant inhabitant, and skulls of purebred Barbary lions from North Africa dating to both the thirteenth and fifteenth centuries have been excavated from the site. By the eighteenth century visitors were permitted entry to the menagerie for a shilling. For those without sufficient funds, however, there was another way to gain entry: it was possible to 'pay' the keepers with a live dog or cat, to be fed, quite literally, to the lions. One gentleman who successfully swapped a pooch for entry found a stray black spaniel, and promptly delivered it to the keeper. Keepers would sometimes starve the lions and other predators for days to provoke a ferocious display for the paying crowds, who wanted to see something untamed and brutish. On this particular occasion, to the disappointment of the crowd but fortunately for the poor dog, the lion, first regarding it with suspicion, went on to form a strong bond with the spaniel. When the rightful owner of the dog heard of its fate and went to claim it as having been lost on the streets, he was invited to retrieve it himself for no-one dared enter the cage to try. The story goes that after a year in the cage, the dog sadly died. Attempts were made to retrieve the body and keepers even provided new live dogs to distract the lion. On all occasions the lion reacted violently and aggressively to any that entered and killed all of the other dogs. Eventually, having refused to eat, the lion passed away with its head on the body of the spaniel, and both were interred together. The lion had been collected and displayed to satisfy a curiosity that people had about the wider – and *wilder* – world but had ended its life displaying characteristics people could empathise with, and was afforded a burial normally reserved for the tame.

Travelling menageries would often advertise the docility of the wildest of their animals in order to draw in more customers, and perhaps this illustrates the real crux of the period. Any place that truly retained something outside of human control was deemed to be a waste and its wild was moved in favour of new crops, new trees, new animals and new boundaries. The wild was no longer just of God's creation; it was specifically and utterly under human rule. It had been something controlled and understood by those who could, people who sought to pick it apart and recreate it, tame it and keep it where it was most appropriate. Elsewhere it was something to be feared, taking on an identity drawn from the myths and legends of earlier periods, with werewolves and witches who communed with wild things and diseases that could cause people to go feral.

The landscape park

As we lost our fear of wilderness, we looked to the countryside as a place of nature, of quiet contemplation, culture and religion. It was increasingly difficult to find anywhere truly remote, as we had taken over the last vestiges of wild land and transformed it into something more useful. Instead, owners of great estates needed to create their own, and the character of these places has persisted in our collective imagination of what the British landscape should look like, and has certainly inspired the lyrical quality of wildspeak. When you are travelling around the countryside, look out for bright shiny green leaves against a deciduous wood, or pepperings of pink and purple flowers in spring. If you see this feral escapee, the rhododendron, the chances are you are on the edge of what is or once was a landscape park. Introduced in the mid-eighteenth century from southern Europe, it is now classed as an invasive species and it is an offence to allow it to be grown 'in the wild'. Rhododendron was introduced as cover for game animals, but also as an ornamental and attractive addition to the grand planting schemes that took place from the 1750s onwards in a bid to create beautiful places that appeared to be 'of nature'.

In the medieval period, deer parks represented the larger designed landscapes, although there was some element of design for aesthetic rather than just projection of power. There were productive monastic and royal gardens immediately surrounding residences, carefully plotted and designed. By the time of Elizabeth I's reign, great houses favoured knot gardens and their successors, parterres, which had been developed in France towards the end of the sixteenth century and were adopted across Europe. Parterres compartmentalised gardens into sections of grass and flowerbeds, arranged in intricate patterns that required a high degree of management to maintain. They were self-contained, not extending too far beyond the immediate area of a house, and had no great impact on the wild beyond the plants that might have been imported. The parterre was immediately obvious as a deliberately planted scheme, and in the end was reasonably short-lived, largely falling out of favour in the early eighteenth century. Against an increasingly ordered backdrop, as the countryside was being straightened and sliced into large fields, and with the dawning realisation that there remained very few (if any) untouched places, estate owners turned their attention to naturalising their surroundings, to creating simulations of an idealised wild world, one of beauty and mystery.

The so-called 'wilderness' gardens of the seventeenth century included features that had naturalistic elements, but were often geometrically arranged and acted as places of contemplation and relaxation. Compartments of woodland might be hedged around by plants that were easily shaped, such as box or yew. Unlike the parterre gardens closer to the house, they did not feature flowers or beds. Paths could either fan out in an array or wind through the trees. Towards the end of the eighteenth century this kind of obvious order and geometric design was given up as the idea of the wilderness was increasingly romanticised. People sought to create copies of what they thought was wild, doing away with anything too neat and sometimes even destroying and relocating whole villages in pursuit of their vision.

Many of the most famous landscapes that followed were created by Lancelot 'Capability' Brown, who was paid vast sums of money after the 1740s for his earth-moving, tree-planting and lake-creating schemes. He had a hand in the grounds of Alnwick in Northumberland, Chatsworth in Derbyshire, and Blenheim Palace in Oxfordshire. His gardens used combinations of architecture, planting and earth-moving to create aesthetic views and obscure disagreeable objects.

The whole was designed to look like an untouched, boundless utopia – achieved no matter the cost.

Brown even removed veteran trees if they did not fit his vision, rather than adapting his designs around such features. The natural *look* was more important than the reality. Animals were allowed into certain areas visible from the house, separated from the immediate area by a ha-ha, a ditch with a vertical revetment on the side that made it appear as though the land in view was contiguous, with no boundaries between the garden and beyond. Clumps of trees were arranged at a distance, in among wider farmland beyond the park boundaries to conceptually bring that area under the influence of the house and help it to appear as less ordered. Grottos and ruins were incorporated into designs, often with running water. Based on nymphaea of the classical world (there are examples at both Pompeii and Herculaneum, for example), they were often decorated with shells and crystals. Capturing the light and sound of the water they sometimes mimicked sea caves, and perhaps the sea beyond Britain's shores was still considered by those who weren't sailors as the last unconquered wild. In some places, the grottos represented something of a more exotic nature. Excavations at a grotto site at Newhailes in Scotland have shown it to have been part of a complex series of features that were designed to give the illusion of a volcano about to erupt. Internal flues allowed smoke to rise from the roof, a quartz-pebble path glittered in the light and cascades of water flowed into a pond that is reckoned to have reflected the entire spectacle.[6] Everything was carefully placed to draw the eye, framed to project an image of nature that was entirely constructed by people. We had reimagined the wild; perhaps Lupa's role at Wilton was to inject some of the danger of landscapes past?

It is very easy to imagine that it was only the very richest who could afford to create this version of the wild on their land, and that it was confined to large but contained areas of the countryside. In fact, large numbers of these schemes happened at a smaller scale and have gone unrecorded and unrecognised. While the leading designers left reams of archive material relating to their endeavours – evidence buried in household accounts, letters, maps, plans and legal disputes – most landowners took a do-it-yourself approach. Certainly, they hired labourers to carry out the work, but the designs might have been the product of a visit to a stately home garden or learned through the grapevine or a book. Beyond the most well-known sites, there are remnants of parks and gardens hidden in plain sight all over Britain. Many of their elements require considerable effort to maintain and keep looking the same. Grazing regimes, forest management and path clearance are all key undertakings for the management of space: a path will quickly be lost if it is not used; a hedgerow may creep outwards if not trimmed. Although when a place is abandoned or forgotten it will lose many of its features, there are those that persist. Water features in particular, where streams and rivers were altered or dug into cascades and spring-fed lakes, tend to survive. Even if the water level falls, the effort might be visible in large scrapes in the ground that would have formed ponds for boating or fishing. In some cases, trees or other plants might spread and take over, but, as with the example of rhododendron, they are recognisable if you know what to look for.

At Hafod in Wales, an entire landscape was formed by resident owner Thomas Johnes from trees such as Scots pine and larch, with cleverly placed paths and gorges used to create views, grottos and baths. Hafod was an artificial wilderness, surrounded by what we might consider today to be a wild and rugged landscape, the moorland of the Cambrian Mountains. But just as with other 'wilderness' gardens, Hafod was not created to be a faithful depiction of the wild that might have been clinging on just beyond the park. As we have seen, moorland is a human-created habitat anyway, but to Johnes it was inhospitable, not good for the poor tenants who came with the estate and certainly not the type of natural landscape he had in mind. It was remote, rocky, and the house was in a dire state of repair. It all needed a programme of reinvigoration. After the mansion was lost in the mid-twentieth century, modern forestry practice was adopted and much of Hafod's designed landscape became overgrown. Terracing, fishponds, paths, the open deer park and meadows were hidden underneath a jungle of growth. Once the house was gone, the landscape seemingly lost the glue that held it together. It didn't truly become wild – the plants taking over had all been specifically planted in a landscape that had long since been cleared and grazed of any naturally occurring trees – but we might consider it to have gone feral. Today, following archaeological appraisal and financial grants, many features have been reinstated and the estate is open to the public.[7]

There are hundreds, perhaps thousands, of these feral landscapes dotted around Britain, places that have lost their houses, their *raison d'être*, ceasing to be managed as faux wilderness parks, and sometimes now simply considered to be particularly attractive 'wild' spaces because their history has been forgotten and things have been allowed to grow unchecked. The reason they are so attractive is because they were specifically designed to be that way. Sometimes it's a single clump of trees in an arable field that is now kept for its timber; at other times it is a bend in the road where trackways were diverted around the edge of a park. The traces linger and affect how we and everything else moves and lives in that space, but it is important to note that just because it was created as a faux version of an imagined wild, it doesn't mean that wild things don't and didn't live there. And just because something has artificial roots does not mean that it doesn't have value. In so many cases the animals and plants that can be found in our countryside are there or able to thrive only because of certain historical circumstances.

Hunters and poachers

Moorlands have long been associated with the wild as a beautiful but imposing landscape where fortunes can change with the weather and animals roam free. One enduring image shared between Exmoor, an upland divided between Somerset and Devon in England, and the Scottish Highlands is that of the red stag, often silhouetted against the heathland landscape or forest. This, our largest surviving large mammal in Britain, appears on branding for national parks, shortbread, tractors and beer. It is synonymous with wild countryside, and the creatures themselves draw in tourists, whether hunter or wildlife photographer. In 2010, there was uproar when the famed Emperor of Exmoor, the biggest stag seen in the south of England, disappeared after allegedly being shot. There were rumours of poachers and trophy hunters, all eager to get their hands on the head of the great deer, which was worth thousands of pounds. Newspapers reported the owners of a local pub receiving death threats for displaying the head of a deer that resembled the Emperor. He had been killed before his time, reports said, and locals were upset that rutting season had been interrupted and he was unable to pass on his genes to a new generation. In fact, for all the articles and outrage, there was never any proof that the Emperor was killed at all. Now, over a decade later, it is highly likely that he has or would have been culled anyway, in the interests of population management, but the story itself has found a place in the myths and legends of the British wilderness.

There are some interesting implications in this story. The deer of Exmoor and the Scottish Highlands are both managed populations. Humans are the apex predator and, in the absence of other ecological controls, we have a role to play in keeping a population healthy, as well as at a level in balance with the surrounding animals and plant life. But the Exmoor deer, held up so often as a bastion of the wild in the countryside, does not appear to be descended from a native population. A study in the 1970s showed that only the Scottish subspecies of red deer seem to have an uninterrupted lineage back to truly native populations, while other deer present across Britain are descended from imported park deer from mainland

Europe that have feralised.[8] Red deer had been kept in forests and parks, as had fallow deer, and although there had been native populations, loss of habitat and high levels of hunting and poaching through time had seen the population dwindle to almost nothing across most of England and Wales by the eighteenth century. At one point there were no red deer on Exmoor at all. In the famous New Forest in 1670, there were only 375 red deer recorded in its entirety, and today the population is maintained at fewer than 100.[9] Deer, in large numbers, can be detrimental to crops, trees and other livestock (the things that people relied on to survive), so while hunting was for the venison and prestige, poaching was sometimes more of a reaction to the destructive impact of deer.

After its peak as a royal pursuit during the medieval period, by 1550 hunting was changing. Rather than the long chase that had been conducted across the largely unenclosed royal forests, the hunt began to morph into a spectator sport. At places such as Lodge Park in Gloucestershire there were long courses laid out with grandstands along their length. A deer was caught in nets prior to the event and carted to the venue, released and chased by sighthounds. The deer was not always killed and the hunt was certainly not always about catching it: on some courses the deer was provided with an escape route. This may have been because already-dwindling deer numbers meant there was little sense in killing every deer put up for the chase, but also because the enjoyment ceased to be derived from how skilled the hunter was and was more about the competition between the dogs. By extension this meant that it was more a contest between those who were training the dogs and less about the nobility or status of the quarry animal. With fewer deer to hunt, less space in the face of an increasingly enclosed landscape and more power to build new things on estates, the gaze of the gentry and nobility transferred from large quarry to small. The focus of hunting shifted onto other wild creatures: first the hare and, by the end of the eighteenth century, the fox. Rather than decimate wild populations, hunting, at least in some places, artificially raised the numbers of animals like foxes and hares with the provision of bespoke environments away from parks and out in farm- and common land. This would have a knock-on effect on how people experienced space.

Hunting of particular animals had been a noble privilege for hundreds of years, but in 1671 the laws governing it were overhauled and became increasingly absurd. Because game, such as hare, partridge and pheasant, was always classified as wild, it could not be owned as property. No-one could claim an exclusive right to the animals that appeared in their fields, parks or gardens, except the legislation upped monetary requirements so that only those who had an income of over £100 a year from owning land had the right and were 'qualified' to take game or own a greyhound. Like older hunting laws that were intended to prevent peasants gathering to revolt, the new law was similarly in part a response to civil war, designed as an insult to the non-landed mercantile class, who were fast accumulating wealth and notions of power.

Not only were penalties for poaching made more severe by the new laws, but over the next century and a half the legislation became so convoluted it effectively meant anyone could be found guilty of poaching if they were in the wrong place at the wrong time. Even more bizarrely, one could be a poacher *and* a gamekeeper at the same time, vested as gamekeeper for one estate but unqualified to take game anywhere else by virtue of being landless. Then there were the poachers-turned-gamekeepers, who, as it turns out, were a very real thing. Job Greenaway was born on 27 December 1767 and spent his life in and around the parishes of Preshute and Manton, near Marlborough, Wiltshire. These parishes are very close to Savernake, a huge and ancient forest that belonged to the estate of the Earl of Ailesbury. The estate was extremely well known for its game. In 1792 Greenaway was convicted of attacking one of the earl's employees, George Newman, beating him and causing various injuries. He was convicted a few years later in the same place for being in possession of a hare. That Newman was involved in such a confrontation is not surprising, as he also crops up in other archive papers as the estate employee responsible for searching tenants' houses for dead game and hanging their dogs. A year after Greenaway's second conviction, in 1797, the earl received a letter headed with the phrase '*Set a Thief to catch a Thief*'. The local gentlemen who paid for the pleasure of coursing on the Marlborough Downs near the earl's hare warren were requesting an extra gamekeeper and asked if they could employ Job Greenaway. Greenaway, they argued, promised that he had seen the error of his ways and if he was employed swore to be 'honest, faithful, and attentive' to his post. I can't say whether he got the job, but I do know that Greenaway was convicted yet again, in 1802, for poaching – within the very same manor he would have been employed to protect. It seems that our poacher was undeterred, for he features in documents one last time in 1820, when he was found in a local coppice wood, probably on a poaching excursion, dead after a 'visitation by God'.

Although the gentlemen petitioning Ailesbury were clearly paying for the privilege to hunt on his land, one significant fact of the new game laws was that if someone was qualified to hunt at all, they were technically allowed to hunt on *anyone's* land. As Douglas Hay once pointed out, wild game had become the property of an entire class.[10] In practice, it was rare for the qualified sportsman to stray onto another's land without permission because they could be prosecuted under civil law for trespassing, but the idea that gentlemen could go anywhere they liked while the poor were excluded from certain places reflected prevailing attitudes. The laws provided an opportunity for estate owners to build new warrens and game enclosures within their landscape parks, but crucially also out on open grass and downland; and because the unqualified could be convicted for simply being too close to them, the presence of these new little parks rendered whole landscapes out of bounds. This usurped the ancient rights of common that local communities had to graze livestock or take wood, and were in many cases a prelude to the large-scale fencing-in of new and very private fields. It was a pernicious power play from a ruling class that used wild things as pawns to commodify what was left of a once open and unbounded set of islands.

The proliferation of small-animal hunting meant adjustments and inventions to allow this to happen, all over Britain. We borrowed duck decoys from the Netherlands, watery places of deception to lure and herd ducks into cages with specially bred and trained dogs, from which they could be harvested.[11] The word 'decoy' itself is said to have entered the English language in the seventeenth century, derived from the Dutch word *kooi* (itself having the same meaning as decoy). While it is unremarkable that Britain contained so many features that were directly to do with killing wild animals, it is the extent to which hunting became a material consideration across the islands, the depth to which it penetrated people's daily lives whether peasant or nobility, that makes our legislated landscape more unusual. Perhaps it is because we are an island nation, but the landscape is one of intersections and overlaps, constantly fraught with competing ambitions, desires and populations.

As ever, woodlands were a central concern in the countryside. In fact, we didn't have enough of them left now to function in the way we required, and thus a great deal of the woodland we see today was, at some point, deliberately planted or managed especially to shelter and preserve animals for hunting. There were pheasantries, strips of woodland that were as useful for their timber and aesthetic as they were for housing pheasants. Pheasants seem to have been kept or at least occasionally eaten in the Roman period in Britain. With introductions of ring-necked pheasants from eastern China in the eighteenth century and enclosures acting as rearing farms, their numbers increased dramatically. Nowadays no-one seems to know whether they are 'wild' or 'livestock': they are classed as wild game so that people can shoot them, but as livestock when it comes to protecting them in rearing pens. This has clear parallels with the confusing status of game animals in the past: wild, but only when it suited certain people.

Other woods were cut through with paths and fenced at the edges to allow for the rounding up of game birds to be taken elsewhere for the shoot. Hare warrens were also wooded, providing hares with a place to shelter in an otherwise open landscape. The brown hare continued to be hunted with sighthounds, necessitating access to big open spaces where the dogs could see the victim for the chase. But these spaces required somewhere for the hare to go when it needed shelter, something to keep it from straying out of the ideal hunting area. Traps were used in gaps in the warren walls and pales to catch hares for release on coursing days. The treatment of the hare's cousin, the rabbit, was very different. By this point, rabbits had been intensively farmed for hundreds of years, and it was during the eighteenth century that they fully naturalised in the British landscape and were able to survive without the pillow mounds and warrens they had been farmed within in the medieval period. They were now as much part of our landscape as other imports such as the hare and the fallow deer, though far from revered.

This is because all this preservation of animals had its downsides, even for the gentry who directed the exercise. Conserving game meant there was more of it to poach and thus there were many convictions for taking or possessing game. These prosecutions were expensive, and associations were set up so that the gentry could pay a subscription for insurance against the cost of convicting poachers on their land. Sometimes landowners also acted as local magistrates and searched houses, confiscating and murdering dogs, setting man-traps and sending spies to lie in wait. The business of protecting these wild animals was a full-time job for many people. A more pertinent downside for many was that encouraging so many animals into one area could be seriously detrimental to an estate's crop yields, as well as to the gardens and small plots of many local people. In 1752 Zeb Lovemore wrote to the Earl of Onslow to complain about the Prince of Wales's hare warren near Epsom, which had apparently devalued a nearby tenant farm because the hares were so numerous and damaging to the fields that few were willing to take it on.

All of this meant that more animals were being artificially preserved than ever before. New trees, hedgerows, walls and drainage systems appeared in the countryside, while the same features were swept away inside new landscape parks to create a vision of a natural idyll. Deer populations that had previously been hunted or hounded to local extinction were propped up with imported stock. Maintaining privilege while also making money was difficult to balance and there was an ebb and flow of habitats and population numbers of different animals as it became ever clearer that our islands were now so heavily managed that any introduction or removal could have a huge impact on what the world looked like and how we lived. We corralled, controlled, investigated and shaped both wild and domestic animals alike. Not only this, but we eroded rights and removed many of the naturally occurring things that would have helped people living in rurality. Correspondingly, the countryside morphed into its new aesthetic.

Enclosing the fields

Standing on most of the hilltops in Britain, unless you are in the middle of the moors or mountains, you will probably be looking at a network of roads, hedges, fences, ditches and walls. The landscape will be a largely irregular patchwork of different farming regimes. Many fields will be arable, growing crops such as rapeseed, which flashes yellow in spring, or the barley that in a summer's breeze ripples like the sea. At various times they will show off the underlying soil type when ploughed: terracotta-red or even blue hues for clay, or silver like birch bark in places where chalk dominates. There will be grazing land and sports pitches, occasional large private or public parks, barns, villages, power lines and factories. Trees might be in woodlands that have had chunks taken out of them over time to form new fields, or they could be arranged in long strips, forming shelter for crops or animals. Every piece of land you see will be accounted for somewhere, owned as property by someone, and very few others will have rights over any of it. Most people cannot access it, to walk, forage or live; only a few birds and smaller mammals and insects will pass through unhindered. There are signs warning against trespass, barbed wire, CCTV and guard dogs. These are private places, largely for profit. If there is one thing that has had the biggest impact on what we see today in the countryside, on how we can (and mostly cannot) access it and what we think of when we imagine rural Britain, it must be the enclosure that transformed it.

Enclosure was the act of redistributing rights into the hands of the few. It normally also meant physically enclosing fields which had often, but not always or everywhere, been open – in the sense that there were no deliberately placed physical barriers to their access. Enclosure was undertaken more rapidly than ever during the eighteenth and early nineteenth centuries, often with the support of parliamentary acts. Land that was regarded as 'waste', not used specifically as pasture or for growing crops, was brought into productive use. This saw open grass downland and meadows filled with wildflowers, herbs and ground-nesting birds ploughed up and their ecosystems destroyed. Old roads were diverted or extinguished entirely, and much common land was lost.

All this had been made possible by two key happenings during Henry VIII's reign, as he wrestled to gain control of religion and land. The first was the dissolution of the monasteries, which effectively disbanded large landowning religious institutions such as monasteries and convents in England, Wales and Ireland (with Scotland following within two decades) and reverted their land and huge incomes to the Crown. This was made possible by the Act of Supremacy, which separated the Church from papal rule and instated Henry as its head. As more land came under the control of the

Crown it was sold off to raise funds for military campaigns. This dispersal of land scattered Britain's landscape into the hands of more people than ever before. The second element worked in tandem with the dissolution. In 1535 the Crown attempted to raise yet more money by closing loopholes that enabled some landholders to avoid paying fees when land changed hands. The new statute effectively meant that rather than people holding the land at the will of the king and having the right to use it, they now owned the full title of the land – just as we are considered to own our houses and gardens now. Shortly afterwards another statute enshrined the right to pass land on to whomever was named within a written will and testament. Not only did we end up with more landowners than ever, but they now had more *rights* than ever, allowing them to manage land how they saw fit, without requiring permission from the monarch to dispose of or enclose it.

Trees and hedge-planting were material considerations in the new landscape and the right types were needed for particular jobs, whether as barriers to access or to provide value in and of themselves. In a landscape where profit was key, true veteran trees would not have been particularly numerous because they were too useful as timber, and so those that survive today that pre-date the sixteenth century really are exceptional and were likely allowed to grow for a specific purpose. Woodlands themselves could be valued simply for their timber – oak for building homes and ships; hazel coppice could be turned into fuel or poles – with a higher value in places that were less likely to support large areas of woodland. There were specific criteria for trees used in the construction of buildings and ships, and in fact many of those grown in woodlands had become tall and straight when fighting for the light, thus lacking the curves required and being unsuitable. It was usually in more open environments that suitable trees could be found: in parks where lone trees had been allowed to mature, or perhaps in hedgerows, where appropriate. Despite increasing needs, the value of trees was constantly being weighed up against other endeavours, just as the value of the deer park had been weighed against the value of the land for arable farming. In a woody place like Sussex hedgerows that were too wide might be cut back to increase the size of the field within. In contrast, where trees were thinner on the ground, estates deliberately planted strips of woodland to create a microclimate in the adjacent field or provide shelter for game birds and from the prevailing winds.

In areas where cattle were more common, new fields needed higher hedgerows than in areas where there were more sheep. The roots of thirsty trees in arable areas might compete with crops and so were less desirable. The hedges of old were normally mixed species and provided forage and fuel from fruit trees, brushwood and the odd pollard veteran tree, but *new* hedgerows were uniform in character and frequently cut to keep them so. Hawthorn was the dominant species, accompanied by blackthorn. If you have ever found yourself clambering over a stile through a thorny overgrown hedge, you will be aware that these plants were the barbed wire of their day. They were there to keep stock animals in and people out.

All of the different requirements of the various boundaries meant a proliferation of different landscape characters, made up of various farming regimes, ecologies, woodlands and settlement types. Old, intimate fields in areas where smallholding predominated, like the North Pennines or western Ireland, had big, stony banks or walls and shrubby hedges, a different world to the large, open arable fields separated by watery ditches in flatter places like Lincolnshire. By 1750 around half of Britain had been hemmed in by fences, hedges, ditches or walls, a process that had taken over 5,000 years. It then took less than a century to enclose another quarter, with each enclosure undertaken by an Act of Parliament, driven by agricultural improvements, rising food demand and the growing power of a landowning elite to convert common land into private property. Now, less than 10 per cent of our land is unbounded, but all is owned and sometimes fiercely defended.

Enclosure by any means had a knock-on effect on the plant life within the field itself. Today, for example, we might pass fields that appear as a sunburst of yellow when buttercups are in flower. They are beautiful, and every bit the picture of rural idyll in the British countryside, but the reason buttercups are able to dominate such a field is normally because that field has been overgrazed. In an open landscape where animals are moving around it is much easier to practise good pasture management and neither overstress the grass nor allow one species to completely take over. Buttercups would be present in a traditional meadow, but their growth is suppressed and controlled by grasses that block out some of the light. Grazing *can* be beneficial, of course: grazing at the right time means that seeds are trampled into the soil and thick swards of grass that might otherwise cover seedlings and block the light are limited. In a traditional meadow we might expect to see other plants such as knapweed, meadowsweet, cranesbill and vetches. In an overgrazed field, however, there will be short grass, dockweed, buttercups and a heavier presence of ragwort. These plants are beneficial to insects, but buttercups are detrimental to horses and cattle alike, and ragwort is poisonous to horses. Still, we persist in grazing our animals in the only land available, perhaps poisoning the unwanted plants but also never seeing the rich diversity of flora that might otherwise appear. More recent developments in pasture management have seen a growth in very rich grass fields that are suitable only for cattle: sugar-rich grass that allows them to grow quickly but which can cause crippling illness in horses, particularly 'native' ones. Improvement of the natural world seems to have been a case of one step forward and three steps back. So many of the fields you see when you stand on that hill will have been seeded for a particular job. Not even our grass is the *real* wild.

By 1800 we had enclosed most of our fields, privatised our woodlands, lost almost all of our megafauna and controlled every inch of the landscape. With nothing to fear in our own countryside and having forgotten what it meant to be natural or wild, we remade our own version in microcosm across the land as somewhere to be enjoyed in private. From Lupa and the lion of the Tower to the landscape park, the wild had become a fiction. All of this seeded what was to come next: a period of ever-diverging beliefs about nature that saw us continuing to affect it with our actions, while we also noticed that the world around us was changing and sought to find ways to minimise the harm we now recognised was being done.

6

PROTECTION

(1800–PRESENT)

> Smoke lowering down from the chimney-pots, making a soft black drizzle, with flakes of soot in it as big as full-grown snowflakes – gone into mourning, one might imagine, for the death of the sun.
>
> Charles Dickens, *Bleak House*

There were spectres on the horizon in the nineteenth-century landscapes of Britain. Thick, dark smoke clogged the skies above industrial cities, and in London the sheer number of people burning coal in their hearths choked the air that found its way into the population's lungs. Contemporary literature was awash with references to this congestion and people understood the cause to be the burning of fuel, but it was simply a necessary part of daily life. The heavy air was the cause of anxiety, but not so much for its effects on people as the fact that it represented the pace of change in a society that was newly industrialised, to which some were resistant. A little ironically, anxiety was also fed by a much deeper worry: that the coal might simply run out. A century and a half later, we'd recognise this pollution as a public health issue and also come to understand its contribution to our now warming climate. For the Victorians, the idea that the climate itself had fluctuated over time was beginning to materialise, ultimately leading to the once-absurd theory that people themselves could be the engineers of that change. At the same time, scientific developments led to a burgeoning understanding of the interplay between solar radiation and the atmosphere, and come the eve of the twentieth century the chemist Svante Arrhenius published calculations that suggested atmospheric carbon dioxide levels were directly linked to the earth's surface temperature. While many were sceptical of this link, Arrhenius seems to have believed that deliberately raising CO_2 levels would actually be beneficial because it would mean future generations could live in a warmer and more comfortable world. He was correct that the world would get warmer, but very few of us would consider this a positive now; perhaps at the time it was an attractive proposition for someone used to long Swedish winters.

While we've touched on the potentially early origins of ecological grief, possibly in evidence dating as far back as the Bronze Age, its meaning was not encapsulated until more recent years. In fact, ecology itself as a term and a concept first emerged only in 1866, when a German biologist and naturalist, Ernst Haeckel, published ideas pertaining to the study of organisms in context. That is to say, he recognised that the natural world is woven from a set of relationships and can be understood by studying those links. While these ideas developed, the actual rate of change wrought by humans on these relationships was becoming more and more obvious, and it is against this backdrop that our approaches and attitudes to what is valuable in a wild world were shaped. This may be why we began to focus on loss and destruction, seeing beauty only in those bits that were most unaffected by us, preventing us from recognising the human element of landscapes we valued and leading us to seek beauty only in what we believed was pure.

The nineteenth and twentieth centuries were a time when we had more global impact than ever, but we also sought to understand our own cultural identity and came to value our rural countryside for its own sake. In the following pages we will look at the developing longing for the past, and how for the first time we turned to think about protecting what we had, not just for the rich but for all. Because we had now begun to grasp that animals and plants are part of a wider set of interdependent relationships, and more and more of us also mulled the experiences of things other-than-human, we were spurred into taking action on what we came to see as transgressions on the natural world. The nineteenth century is when 'sports' such as bear-baiting were banned and the Royal Society for the Prevention of Cruelty to Animals (RSPCA) was founded, the first dog shows and animal parades took place, and the first zoos were opened in this country. Environmental protection was slower to appear: the twentieth century saw the advent of national parks in England and Wales. These were partly influenced by a similar movement in the United States which sought to preserve tracts of wilderness, but here the driver was primarily protecting access and landscape beauty in the face of development. That 'beauty' was not necessarily wild and untamed, but also included farmed landscapes which were integral to rural character, and as a result it failed to reckon with some of the detrimental environmental impacts of modern farming practices. The railways connected people to these landscapes and fed a nostalgia for days gone by against a backdrop of rapidly accelerating change. By the time we arrived in the late twentieth century, 'wild' meant anything that lived in 'nature', and 'nature' was anything green and fruitful and perceived to be operating independently of us.

Longing for the past

In the early nineteenth century, while the Grand Tour of Europe was still a popular way of educating the young men of Britain, attention had also begun to turn inwards and backwards, to draw on the beauty and inherent *native* qualities of the British landscape. The artistic Picturesque movement framed, literally, the landscape as a harmonious whole made up of undulating as well as jagged parts, where waterfalls cascaded off rough mountains, forming a backdrop to the curve of a weeping willow reflected in a river or lake. As it reached its peak the movement comprised not just visual art but writing: Wordsworth famously declared the Lake District to be an area that the entire nation had the right to enjoy, there for whoever 'had an eye to perceive and a heart to enjoy'. He consistently praised its wildness, but also said that Gowbarrow Park, a medieval deer park on the hills, was where 'a lover of Nature might linger for hours', with its brook dashing through a deep glen, ferns, hawthorn and honeysuckle lining its steep sides, and fallow deer 'glancing and bounding over the lawns and through the thickets'.[1] For proponents of the Picturesque the past was a very real component of a beautiful landscape: often this took the form of a ruin sat among the bushes and trees, sometimes being slowly consumed by natural creeping predators like ivy. Such paintings give off the kind of 'nature is healing' vibes we see cropping up in social media posts now, depicting the greening of formerly 'human' spaces, but they were also part of our cultural identity – a romantic wild beauty that made Britain *Britain*.

The medieval period was drawn upon as a golden age, replete with valorous knights and mighty castles surrounded by forests and common fields aplenty. Classical architecture seen on travels through Italy and Greece remained popular. The 'wild places' were more than ever a fiction or mystery and some places were admired as naturally occurring beauties that had the capacity to recover from human intervention. It was an early appreciation of what we think of as rewilding, perhaps, but certainly also a longing. A word for this, coined by John Koenig, is 'anemoia', a longing for a time one has never experienced, which for the Victorians was also a past in which the glory days could be found. This anemoia for a wild and untamed past has stayed with us. It's frequently visible in the way we talk about nature now: if we could only go *back*, what peace and satisfaction we would find in a world in balance. I can't help feeling that we're actually hankering after an imagined past that has never *really* existed. There is also a term for this, one of those wonderful words that has no direct equivalent in English because we haven't thought to articulate it. The term is *hiraeth* in Welsh, or *hireth* in Cornish, and it is loaded with such complexity that it's difficult to describe. It is a deeply felt sense of homesickness, a longing for somewhere you can never return to or that might never have even existed. This, for me, is what was sown by the writing about wild things in the nineteenth century, the thing that rooted itself in our very sense of what rural places are and should be and drives us to seek out a kind of Eden in which we can immerse ourselves in everything that is *natural.* It took root easily during a time when more and more people were living urban and suburban lives, in air that was thick with soot as technological innovation gathered pace. It's evident now in all the photos of natural beauty hotspots on social media, with carefully positioned cameras that project a representation of an ideal, but with scores of other tourists waiting for their turn just out of frame. We long for a wild and a past we've never experienced and has arguably never existed, which ultimately blinds us to the wonder and beauty in the things we still have.

Contaminated wilds

Some way up the winding river from Plymouth sits Endsleigh Cottage, deep in the woods on the Devon bank of the River Tamar. It was built in the 1810s in the style of a 'cottage', as a retreat for the Duke of Bedford and his family. Humphry Repton, famed landscape gardener and often regarded as Capability Brown's successor, reworked the surroundings into a pleasure garden in Picturesque style, complete with a 'pinetum', a coniferous arboretum, and a Swiss-style cottage set in an alpine garden. Its wider landscape featured a network of carriage drives, particularly along the river, along which were dotted small buildings for taking tea. The Tamar Valley epitomised the Picturesque in many ways, with rocky granite outcrops, wooded valley sides and a wide tidal river, and grew to become a popular Victorian tourist spot. People would picnic on the aptly named Chimney Rock, catching a boat up the river from Plymouth at its mouth. It was seen as the perfect exemplar of natural beauty, so much so that it was immortalised in the work of J. M. W. Turner, who sketched and exhibited it in a painting called *Crossing the Brook* in 1815. It currently hangs in Tate Britain in London and features the river front and centre, leading away between steep valley sides in the distance. At first glance it seems to be a painting mostly of nature, but there are telling signs that there were other things going on. To the right of the main river scene is the gaping mouth of what is easily mistaken for a cave, but look more closely and you will spot a flat top to its apparent face – it's a lime kiln, used for burning lime to produce fertiliser to ship up and down the river. Downriver is a bridge, the most southerly on the river until Isambard Kingdom Brunel built the Royal Albert Bridge at Saltash in 1859, though it was only for the railway. But other buildings – a mill next to the bridge, and chimneys dotted among the trees – belie the increasingly industrial nature of the valley that was presented as the epitome of nature's beauty.

There had been mining in and around the area for centuries: there is a Roman fort nearby and it is likely they exploited the local resource, and there was also extensive medieval silver mining. From the early–mid-nineteenth century, however, the river was to become a core part of the mechanism by which huge amounts of copper, tin and arsenic were taken out of Devon and Cornwall. Parts of Endsleigh were built to take advantage of a view towards Kit Hill on the Cornish side of the valley, which was – and still is – topped by a tall chimney that acted as an eyecatcher, not for what it was, but because it was shaped as a classical column and could therefore appear like an ancient ruin. As time marched on, it cannot have been easy to turn a blind eye to the sheer amount of industry that was happening all around the estate. The Duke's Drive passed through or close to multiple workings, all with infrastructure and dozens of shafts and adits deep into the ground. The valley also proved something of a microclimate, and woodland was cleared for market gardens which developed their own strains of early strawberries and daffodils, sent up to London on the railways that had been built to move the geological wealth coming out of the ground.

Ironically, the mines, now long since abandoned, have been 'reclaimed' by nature in a way that might have been proudly showcased in a Turner. The area is part of a World Heritage Site but is also part of a National Landscape, formerly known as an Area of Outstanding Natural Beauty. For a few years, we lived in the valley while my job was to connect people to its landscape heritage. It really is beautiful, but I'm not sure how natural we can consider most of this place that has been stuffed with very real and very invasive activity for so long. Some of the nature that has reclaimed the river is not welcome: there are Himalayan balsam and Japanese knotweed in abundance, both introduced for their beauty in 1839 and 1850 respectively, and now having significant impacts on the natural (and in the case of knotweed, the built) environment. Balsam hitches a ride up and down the river, smothering other plants on the banks, but is fairly easy to destroy by simply cutting its translucent hollow stems. These unusual stems and its large pink flowers make it quite a sight, and it's instantly recognisable when you know it shouldn't be there. Knotweed is notoriously difficult to remove, as it is rhizomatous and chopping it up will mean you've just propagated thousands of new plants with renewed vigour, like ground elder on steroids. The most effective method of control is with glyphosate, which will kill other plants and insects in the area. In Japan there are natural controls – a type of insect as well as a fungus that can kill knotweed – but so far, there doesn't seem to be an effective biological control method that we can use here and trust not to take out a whole load of other plants. We make decisions like this in the present because we know they didn't in the past; we now understand the butterfly effect of making spurious introductions into established ecosystems. Owing to all of the industrial activity, the land is also deeply contaminated. A recent study of a leisure park that has been built around some of the old workings found dangerous levels of arsenic in the soil and dust that is kicked up on a dry day by walkers and bikers, even though remedial works were undertaken some years ago.[2]

Despite the negative aspects of the industrial workings that leech into the present, this particular blend of historical activity with the natural environment has led to unique opportunities for some animals and plants, providing habitats that draw in species that otherwise wouldn't choose or be able to make this their home. The endangered greater horseshoe bat would ordinarily roost in caves but here roosts in the abandoned tunnels of mines that dot the scarps. There's also the nightjar, a nocturnal summer visitor that is confined to the peripheries of Britain, with almost none in Ireland and very few in Scotland. Its Latin name is *Caprimulgus europaeus*, the first half of which means 'goat-sucker', because it was widely believed to steal milk from goats. This myth was born of its silent flight, hunting insects, which often cluster around livestock, in the gloaming – it is fitting that one of the nightjar's favourite foods is said to be the ghost moth. Because of this, they are usually found in mosaic habitats of woods, farmland and orchards – the kind of mosaics made by people. Its unusual name has survived because it was chosen in the eighteenth century as a formal label, part of the classification of things we covered in the last chapter. Were it not for that, perhaps it would have been lost. Another rarity to make its home around the old mines of the Tamar is the heath fritillary butterfly, one of our most scarce. It is a successional species, so relies on continuous change and the provision of new woodland clearings with young trees. For this reason, it has been particularly associated over time with coppicing, earning the nickname 'the woodsman's follower'. It is a relationship that must go back centuries, but has now been jeopardised because we just don't manage woodland in the same way. Instead, we need to replicate the successional process on which the butterfly relies in order to ensure its future, looking to the past to understand how to protect its future.

The Tamar Valley is often described as unspoilt, which is normally taken to mean that it is not blighted by human development. There clearly isn't really an absence of human activity, though; it's more that vegetation has grown to largely conceal from the immediate view what would be considered to 'spoil' it. More importantly, without its specific cultural history, many of the rare species that make it such a special place wouldn't even be there. We've tended to recognise this with farming, but not necessarily with other places that we would consider 'brownfield', those that have developed in the past. Not all of these places have much ecological value, but some have been shown to be just as biodiverse as habitats that have been undergoing development through the interplay between people, crops and animals over centuries.[3] We need to replicate the activities that these species have adapted to in order to ensure their survival – they are able to thrive *because* of the ecologies that have come into being with our presence over time, however unintentional.

Protecting the land

Wordsworth's writing on the beauty of the Lakes is often described as the first nod towards the idea of conservation in the landscape, though it might be better described as *preservation*. These early calls were about protecting things as they were without necessarily thinking about the processes that had made them in the first place. It was a kind of halting in place, designed to be looked upon, but not necessarily by the masses who might be inclined to ruin it. However, through the latter half of the nineteenth century several societies and groups formed that sought to protect areas not just for their aesthetic value, but to provide access to green spaces, particularly for the working classes who had been dispossessed of so much land by successive enclosures and the prevention of access. In Scotland, the notorious Highland Clearances had removed huge numbers of upland tenants in favour of a consolidation of land among the few, as well as use of the land for sheep, for instance. Among the early groups were the Commons Preservation Society, now the Open Spaces Society, which was the first British national conservation society, formed in 1865. This aimed to protect commons, village greens and, later, footpaths. On their committee was Octavia Hill, who went on to drive the formation of the National Trust in 1895. The Trust had twin aims: to conserve the best and most *beautiful* parts of Britain (Scotland was to get its own trust in 1931, and Ireland in 1946), but also to maintain access to these for the poor.

The fact that rising populations in cities were hampering the fitness of entire communities, who were growing up small of stature and with 'narrow chests' and a 'generally unhealthy appearance', was seen as a matter of public health concern.[4] Cities were bad for people, or at least bad for what people supposed was a civilised society. The countryside was transformed into something that could provide utility in terms of leisure and aesthetic, a place to escape from the industrialised cities that were becoming ever more clogged with foul air. The new trend for visiting places away from home brought with it conflict – not with the countryside itself, as a place to be wary of dangerous predators, but with other people, over the right to be in a space. As with the rationalisation of fields during enclosure, there had been attempts in the eighteenth century to do similar with the highways and footpaths, in England at least, but it wasn't until the nineteenth century that it really became a wider issue. One of the first proper battlegrounds was the River Clyde outside Glasgow in Scotland. Landowners seem to have been attempting to exclude the public at large from a favoured walk along the river for years, but in 1822 people began to organise to oppose this loss of liberty.[5] It took years, but eventually enough people testified to the historical and habitual use of such walks along the river that access was restored by right. Similar movements were happening further south, with the first footpath preservation society created in 1824 to prevent the loss of access to areas of the countryside around York. These were islands which, we now recognised, had things that might not last forever, that we might lose access to and that we needed to preserve.

Alongside protecting access to green spaces outside of the city, green spaces were created within it, and the establishment of new public parks, which had first appeared in the early Victorian period, gathered pace. Rather than attempts to create an idealised nature as in Brown's eighteenth-century landscapes, the new parks were organised and categorised forms of nature and some really were more like open-air museums. After the Great Exhibition in 1851, which showcased art, objects and machinery from all over the world, the building that was designed to house it was moved from Hyde Park to the area that became known as Crystal Palace. A park was created around it and within were sited over thirty dinosaur sculptures, based on recently identified fossilised remains – they were the first-ever attempts to create life-sized models of extinct animals, and regarded as the first scientific outreach project. They weren't, of course, completely anatomically correct, because we've since learned a whole lot more about dinosaur mobility and how they might have actually moved and looked, but they were quite the spectacle. The dinosaurs are now Grade I-listed, although they are considered at risk of imminent deterioration and conservation work is ongoing.

Despite the moral drive to provide spaces for the masses lest they degenerate into unhealthy and recalcitrant criminals, legal designation of large tracts of land like the national parks we have now did not arrive until the 1950s. This wave of designations – from the Peak District to the Brecon Beacons – was part of a post-war shift in which access to nature was increasingly seen as a necessary part of rebuilding a healthier and fairer society. It is probably no surprise that there is extensive overlap between what are now national parks and what had been maintained under the old forest law: they were areas where habitat preservation of some form had been happening for centuries, even if it was just to enable hunting. But these were also all uplands, places where settlement had always been less dense, and so they appeared wilder and more natural to the eye even though they were the product of thousands of years of human intervention. They were certainly rugged and jagged and contained mixtures of wood and pasture, moor, heath and plenty of cascades that literature described as the pinnacle of native beauty. In a way, the early national parks were established on the basis of a false narrative. They looked like what we'd been told by romantic writers was wild and natural, a last vestige of somewhere that the advances of human industry had not yet touched and should therefore be preserved in stasis.

Monocultures

With some of the landscape now demarcated and protected, elsewhere there was also a real need for the land to *provide* – it couldn't all be preserved for its beauty, and the practices that proliferated in times of need were far from sustainable. While the mines were at their peak and when enclosure was all but done, there was a period of contraction in domestic arable production as the country came to rely on cheap grain imports from elsewhere. As a result, much land was turned over to sheep and dairy farming, although some also fell out of use, but it meant that less than a century later we found ourselves unable to provide enough food for the nation when trade was interrupted by war. In 1947, the Agriculture Act was instituted to boost homegrown crops, notably encouraging the use of mechanised farming that was suited to very large farms and correspondingly large fields. It also meant more specialisation in specific crops rather than mixed regimes. By the 1960s, there was widespread removal of hedgerows to make bigger and easier-to-manage fields, grants for the grubbing up of orchards that were thought to be incapable of providing financially viable crops, and old meadows were reseeded with improved grassland. The result was a hastening of the decline of mammals, insects, birds and diverse plants. In fact, only amphibians seem to have seen any stabilisation or rise in numbers since the 1980s, thought to be due to an increase in the number of ponds in people's gardens, which counteracted the loss of old dew and other ponds out in the fields.[6] This rise is not uniform, however, and the native common frog seems to be coping better than the native common toad.

These losses have not been driven, in most cases, by the destruction of wild spaces, but by the modification of anthropogenic ones. Hedgerows, arable crops, orchards: all were created by us, and the mix of species in them occurred only because of the type of habitat that had been inadvertently created over hundreds, sometimes thousands, of years. The most successful species left today are habitat generalists that can cope with the increasing homogeneity. Intensification of agriculture therefore threatened species such as the skylark, harvest mouse and corn bunting by altering the availability of vital food sources. But the same thing would happen if we tried to remove agriculture entirely – the animals have come to rely on the way things have been done over time. They are a snapshot, really, a particular mix of things that have moved in to occupy a specific set of conditions. They haven't always been there. Still, we do know that the general trend is one of loss of diversity because our surroundings were themselves becoming less diverse at a rate of knots in the twentieth century.

As Victoria's reign was ending, the timber requirements of the nation already outstripped supply, and when the First World War exposed Britain's vulnerability in this regard, something needed to be done. Suitable places were heavily planted with imported conifer stock, in neat rows, all at the same time, so that they grew as a collection of totems with a dense canopy above, creating the dark, dead space beneath, littered with needles and the occasional fungus. These were the very antithesis of the *natural* pine forest, at that point confined to small areas of the Scottish Highlands, which would have been filled with trees of different ages and included broadleaf species, in a pattern that supports a far more diverse set of entanglements than a few mushrooms. Such plantations are instantly recognisable – they are utterly impenetrable to light, and they frequently sit as a harsh pocket of uniformity within and alongside deciduous woods. Many of these are now coming to the ends of their lives, and when the timber is harvested and the whole is opened up to the light it is surprising just how quickly things regenerate. Suddenly the understory springs to life with all of the plants that have been lying dormant in the darkness.

(Thinking about) Protecting the animals

As anxiety developed about access to space and some habitats were given over to intensive production, there were also drives to protect animals. The RSPCA was established in 1824 and received a royal charter in 1840. It was founded in a coffee house in London by a small group of influential people, among them William Wilberforce, and was principally concerned with the treatment of domesticated animals. In particular, the group sought to push forward legislation that dealt with cattle markets, which its members saw as cruel. Later, it took on bear-baiting and cockfighting, and then widened its net to campaign against things that harmed wild birds. It wasn't, however, particularly concerned with blood sports because many of its members, belonging to a particular class of society, were hunters themselves. The barbaric treatment of animals was seen as a threat posed by the uneducated lower classes, and for that reason the hunting of wild animals just didn't seem to count – bear-baiting and cockfighting were working-class sports and it was for this reason they were targeted. The dirty cities with their pubs and gambling provided too many opportunities for the labouring classes to skip the labouring part. Banning these sports was in line with preventing fourteenth-century peasants from hunting so that they weren't meeting up and plotting revolts. It was not until the 1920s that a group, which became the League Against Cruel Sports in 1938, was formed to deal with sport hunting as an issue of animal cruelty.

But although these groups began to form and certain ways of treating animals were banned, more than ever were being killed. The Victorian countryside was a place where we preserved some animals – expressly to then be killed in a specific way – and persecuted others for being inconvenient. Building on the trend from times before, sport hunting had become a huge part of rural life by the nineteenth century. In a single season on the Elveden Estate in Norfolk, between September 1885 and February 1886, over 80,000 animals were shot for sport. The majority of these were rabbits, but the figure also included around 12,000 pheasants and nearly 2,000 hares. This number seems extraordinarily high, but in fact was not unusual. Game animals, but especially pheasants, partridges, hares and even ducks, were encouraged, nurtured and preserved. They were provided with suitable habitats and encouraged to breed in great numbers, all so that they could then be shot or coursed. Even foxes were preserved to a degree, provided with artificial earths and coverts so that there would be an adequate number to pursue on horseback. The animals were not bred in barns or too tightly enclosed because they needed to retain something of their wild nature to make the hunt more authentic, and so were provided with areas specifically planted and designed to emulate their preferred locales. Since the hunting ban in England the number of foxes has fallen back to what are probably much more natural levels, excepting perhaps a bump in the number of urban foxes as they find new ways to live in our world. The numbers dropped by about half between 1996 and 2016, with numbers falling more seriously after 2008.[7] Ironically, it may be that the centuries of preserving foxes only to kill them could feasibly have prevented them from being persecuted and disappearing entirely sometime in the nineteenth century or earlier.

The fox as vermin, but simultaneously important enough to keep alive, is an example of the kind of conceptual confusion that became attached to many animals. The pheasant, for instance, seems to morph in and out of being wild depending on the situation. It's a non-native introduction and seems to have followed a similar pattern to some of the other animals that first appear in the archaeological record in Britain in the Roman period, with little to suggest there were any present in the early medieval period. A resurgence in the following centuries hints that the pheasant was probably initially a park animal but by the fifteenth century had dispersed into the wider landscape. Today, it is seen as a quintessential game bird, and it's this game status that causes the most confusion. We saw how in the seventeenth and eighteenth centuries 'game' was a legal definition that meant an animal was wild and therefore could not be owned. Now, of course, they're reared on specific game farms and then sold to shoots, where they are eventually released – tens of millions of them every year – into the countryside. When penned they're classed as livestock, but when they're released they become wild. Or do they? While they're designated as wild, no-one is responsible for any damage they may cause, but they do frequently return to areas where they are fed, which is often in their original pens. Does this lend them one reality by day, another at night? During open season, legislation means whoever owns the land they're on can shoot them. At other times, because they are wild they are protected, but critics suggest they're eating their way through our population of adders and slow worms too. On the other hand, woodland management for pheasants is thought to be beneficial to songbirds and butterflies, and the control of predators of the pheasant is beneficial to other species we might want to protect. The pheasant is at once both an icon of Britain's countryside and a symbol of those who control it.

Another member of this club is the badger. Historically used for its fur, it does appear in the archaeological record over time, but it is also revealed in churchwardens' accounts in the early modern period as vermin that people were paid to eradicate, like foxes. Despite this, badgers did not come into conflict with people very much – they weren't stealing anyone's crops like rabbits were. They appear in folk tales and superstition, and it's easy to see why: they are elusive and silent nocturnal animals, but if you come across one you might catch a glimpse of its white stripe through the dark, and if cornered they can become very aggressive. In Irish mythology they are shapeshifters.

Badger-baiting was also a bloodsport that sat alongside cockfighting, bear-baiting and dogfighting. Badgers were dug from their setts and kept in pens, until they were loosed into an arena to be killed by dogs in front of a baying crowd. Badger-baiting was outlawed in 1835, and the badger subsequently took on a new persona, expounded in tales such as the *Wind in the Willows*, where the badger was a gentle and wise old thing. By 1900 they were declared to be rare within Britain, though it's now thought this was false and due to the fact that we hadn't ever really taken the time to understand them before and just weren't aware of how elusive they are. Their true numbers are far more apparent now for several reasons, not least because so many are killed on our roads each year. We've had an increasingly complex relationship with badgers over time, but the real dichotomy became apparent in the twentieth century. This was when the badger of a nostalgic countryside was pitted against a new badger identity, that of disease-ridden killer.

In 1971 the carcass of a badger was discovered on a farm and taken for testing. It was revealed to have died of bovine tuberculosis (bTB), and was an indicator of what was to come. Within a few years there was a full-scale cull of badgers, in an attempt to eradicate a disease that was killing cattle. Badger supporters argued that indiscriminate culling would take out healthy as well as infected individuals, would significantly affect the social structure of badger groups and didn't actually work to keep rates of bTB down: it seemed to temporarily alleviate rates in the immediate area but pushed them up elsewhere. Those who supported the culling seem mostly to have argued that it wasn't just cattle they were protecting but other forms of wildlife, on whom badgers both predated and could pass bTB, but also because they can and will eat other small mammals, including the beloved hedgehog.[8] In fact, both hedgehogs and badgers feature in the perennial favourite illustrated children's book *Brambly Hedge*, a series of eight books released from 1980 onwards that depicted rural animals living a lovely wild life but under assault from familiar villains like the ferret and the rat. Despite this, the badger is not only disliked by cattle farmers but is also often criticised for the damage it can do to gardens. It gets a bad rap in archaeology too: badger setts can significantly affect buried archaeological remains and because of their protection a licence is required for any work that can impact the sett or its inhabitants.

What really characterises our relationship with animals as it developed through the Victorian era and beyond is that they became pawns to be deployed in sport or by both sides of the debates that arose as a result of increasing tensions between the different groups that had all sorts of interests in our wild or not-so-wild countryside. As fierce arguments were made for and against particular animals, still the numbers of new characters in the story rose.

Introductions

In 2005 the *Independent* reported on what it said was the 'silent killer' of Britain's woodlands: a small, 'pig-shaped' animal with 'antlers like salad tools' – the muntjac. The description is fairly accurate: the muntjac is a bit bizarre-looking, and very small, an adult easily mistaken for a dog and its young for a cat. The first muntjacs here were kept on the Duke of Bedford's estate at Woburn, which in the late nineteenth century held a collection of different species of deer. A book written in 1898 by Richard Lydekker on the history of all deer was partly inspired by Woburn's collection. He stated the muntjacs 'were allowed to run wild in the coverts, where they thrive well, although of course they are but seldom seen'.[9]

The turn of the twentieth century was therefore the beginning of the great takeover. Shy though they are, muntjacs have received increasing attention because of their escalating spread. An article in 1958 claimed that the first muntjacs to be released were known as southern red muntjacs, indigenous to much of South East Asia, including India and China. These were aggressive and posed a risk to 'small dogs', and so attempts were made to replace them with the one we have now, Reeves' muntjac. At that time, the author estimated there were only around 500–1,000 muntjacs in Britain, and they were confined to a range, albeit one that was acknowledged to be growing, of around 100km from Woburn itself.[10] The spread was driven not just by the initial 'wild' Woburn individuals but also by subsequent releases. By the 1990s the population was thought to be at 50,000, and now it may be triple that, with a range that covers most of England and Wales, with only a few in Scotland. Beyond this, they have been introduced to Ireland as well as to continental Europe, where there is real concern about what their rapid spread will mean for habitats.[11]

You may wonder why they are such a problem, when Britain has survived so many other types of deer introductions, but the deer population was already an issue. We simply didn't and still don't eat enough of them, and muntjacs do also have some habits that make them a more serious threat. The issue is with how they feed, both browsing saplings but also having significant impacts on woodlands rich in ground flora, which are also more likely to be SSSIs. They'll eat the tastiest grasses and leave the least palatable to proliferate, as well as consume bluebells and oxlips, and in some cases their presence has meant that once widely spread orchids now grow only in protected enclosures. Other than fencing, which is not always practical, the only effective method of controlling them has been culling, which is also the way that other invasives have been dealt with.

Better known for its destruction is the notorious American mink, brought here in the late 1920s to satisfy the demands of the fur trade. Like rabbits had been in the thirteenth century, mink were farmed, but eventually went feral. Unlike rabbits, mink were able to colonise quickly, not needing to adapt to our climate in the same way. Their indigenous equivalents and competitors, the otter and the polecat, had been hunted and persecuted down to the smallest of numbers, and mink more than moved in to take their place. These three animals form an interesting group of similar species with vastly differing statuses – and different consequences for the remaining wild of Britain.

Same family, different fates

The otter is an icon of our waterways, but is rarely seen outside of a zoo. It was a particularly popular quarry animal in the Victorian period and beyond, hunted by dedicated packs of hounds and hunts in a similar vein to the fox, the deer and the hare, with different groups hunting areas of land organised around river systems. Thousands of otters were killed this way, particularly in the first half of the last century, right up until a precipitous drop in population numbers in the late 1950s. Otters had been utilised for their skins since at least the Iron Age, but there don't seem to be a huge number of them occurring archaeologically. However, genetic studies show that prior to the big drop there had been long-standing population pressure for centuries, suggesting that factors like habitat loss and hunting were taking effect in the medieval period.[12] The steep decline in the fifties was certainly in part due to hunting, to which many were opposed, but fisheries, too, trapped and killed otters to protect their stock. A major factor in the decline of otters at this point was also the ubiquitous use of harmful agricultural chemicals that seeped into their watery homes. In recent years, legal protection and environmental mandates on water quality, as well as species reintroduction programmes, have seen them make a comeback and they have become a tourist draw in some places.

Half a century prior to the otter's near-death experience, its smaller relative, the polecat, nearly met its own end. Persecuted by gamekeepers because of its predilection for game animals, by the early twentieth century it was confined to Wales and the borderlands with England. Now, the polecat is the subject of conservation efforts because of its status as an incumbent native and its population has increased, but it is considered under threat from feral populations of ferrets. The story held in its DNA is, however, a little different, showing the polecat is still predominantly confined to Wales – because everywhere else the 'polecat' is actually a hybridised version with significant ferret admixture.[13] The polecat is the wild ancestor of the ferret, which has been domesticated for over 2,000 years and was used in Britain, particularly in the medieval period, for hunting rabbits. In this way the ferret and the polecat have a relationship akin to that of the dog and the wolf. One was created by humans to be used for their own ends; the other is seen as a threat. Then there's the notorious mink, as described above, whose presence has been credited with destroying the habitat of the water vole. So, we have the otter, king of the slippery water mammals, the mink as a dangerous interloper, and the polecat, which can be something in between, all fulfilling different roles in the theatre of wild things but all heavily affected by the nineteenth- and twentieth-century thirst to satisfy hunting as a pastime and trade even as thoughts turned to how the environment might need to be protected.

Reintroductions

Species restocking, reintroduction and culling are fascinating and extremely current topics. The last 200 years have left us with an overabundance of some animals and an underabundance of others, and are complicated by introductions that quickly colonised. Many were introduced for short-term gain, without thinking about what that would mean for the future. This was despite a growing conservation movement that was well aware of the changing dynamics of the landscape and the wild, with people increasingly expressing concern at the over-exploitation of wild animals. Attitudes tended to deal with the immediate – what direct impact were we having on animals and plants, and could we mitigate it by simply stopping? Some animals, like the otter, have recovered because we stopped hunting and poisoning, but others like the fox are now declining *because* they were previously protected. We've tended not to look beyond hunting as a way of killing and consider its knock-on impact. This characterises much of the way we think about the wild – people are bad, and if we just remove all trace of them it'll recover. Early conservation protected things from the impacts of people, but not at a distance.

Releasing animals like the muntjac and the mink could be driven by factors ranging from their aesthetic value to accidental escape. Like the muntjac, the grey squirrel was introduced in the nineteenth century effectively as an ornamental, in much the same way as the balsam and the knotweed. And just like those, it is now considered an invasive: it transmits the deadly squirrel pox to which our native species, the red squirrel, is more susceptible. The narrative around grey squirrels is overwhelmingly negative. They are better able to digest a wider range of foods, including nuts and seeds that have yet to ripen, and there are now estimated to be more than fifteen times the number of grey squirrels than reds, which have refugia in some parts of Scotland, and on places like Brownsea Island. Because of this, we're far more likely to see grey squirrels, particularly in urban environments where they will hop from garden to garden taking food from bird feeders. Their propensity to spread disease and run around the city means we think of them more as vermin, just as we do urban foxes, pigeons or rats. But hundreds of years ago, it seems likely that it was actually the angelic red squirrel that gave us leprosy, via furriers who obtained and processed their skins.[14] Indeed, the Brownsea Island squirrels, a major draw for tourists visiting the island, have leprosy – and have had it for hundreds of years.

What stalks the woods

Other animals were introduced based on a perceived need. In Epping Forest in the 1880s there were reports of strange grey foxes living in the woods and even attacking dogs. One very unusual fox cub in particular had been seen by poachers and labourers late at night or early in the morning, described as having a 'very bad brush' due to its coarse grey coat. He was caught, named Charlie and brought up with a fox-terrier bitch with whom he eventually mated. Charlie could be very aggressive, particularly with other rough-coated dogs, and regularly attacked cats. He developed a sort of loyalty to his owner and was said to be deeply suspicious of strangers. Charlie's puppies were 'nasty spiteful animals, having the bad temper of both parents and the good qualities of neither'.[15] This is because Charlie wasn't a weird kind of fox at all, but a coyote from either North America or North Africa, and he had come to Epping Forest having been sold as a fox cub for hunters to replenish their dwindling prey numbers. Peter Steinhart, in his book *The Company of Wolves*, has pointed out that every time we hybridise a domestic dog with a type of wolf, we are resetting the clock, reintroducing an aggressive nature that has taken us upwards of 10,000 years to attempt to eliminate.[16] It's like almost finishing the build on a house of cards and then trying to replace one of the layers in the middle and being surprised when the rest fall over. Restocking of wild quarry was fairly common, but quite how a litter of coyotes ended up in a bag on someone's cart in Epping is a mystery. They survived, and probably bred, possibly even hybridising with other local dogs, though they never became quite such a legend as the other exotic animal that is often said to stalk the countryside of Britain – the big cat.

Stories of big cats that stalk the moors normally come in the form of folk tales, or maybe along the same lines as UFO sightings – misidentifications in the darkness. Beyond the anecdotal sightings and grainy photos found in local newspapers, there have been some more scientific approaches to understanding these stories. The evidence is admittedly a little sketchy: in 2024, the BBC claimed that DNA on a lamb carcass had been analysed and was shown to belong to a big cat of the jaguar or mountain lion family, although there is no scientific report to back this up.[17] However, one study has used archaeological techniques to examine the skeletons of animals that had been killed in the Cambrian Mountains of West Wales, carefully measuring the marks left behind by the teeth of the killer to ascertain likely size and species. While the purpose of the study was to look at the implications for understanding archaeological collections of animal remains, the results strongly suggested that at least one of the lambs had been killed by a medium-sized cat.[18]

I'm sure many are sceptical, but even this in itself is interesting: we regard the possibility of big cats roaming around with humour, most assuming them to be false identifications. It must be at least in part due to the ease with which we began to venture out into the countryside in the nineteenth century. Presented as a health opportunity, it was also fed by the new railways, which took people out to visit beauty spots, which were homogenised into what was considered natural – out there, if it wasn't a farm it was probably wild. In the nineteenth century there also emerged other places we could visit in order to experience the exotic, which were to become living museums showcasing species from around the world: zoos.

Zoos and safari parks

The first zoos may technically be considered the menageries we've seen evidence of in the Roman world, in the Tower of London and on great estates. The Tower was something of an exception in the way people were allowed to view the creatures; most menageries were simply a way for the gentry and aristocracy to show off their reach and power. The zoo was to embody something else: it was envisaged as a modern scientific institution. Alongside this emerged its role as a way for the masses to encounter wild and exotic creatures, though in common with the menageries it could also be considered a way to showcase imperial dominance.

The first modern zoo in Britain was the Zoological Gardens of London, opening to fellows of the Zoological Society of London

(ZSL) in 1827. It was intended to be a place where living animals could be studied to understand their natures and habits. Many of the animals arrived as gifts from nobles and companies around the world – a beaver from Canada, and a brown bear from Russia. Some animals were of uncertain origin, including a Celebes crested macaque which was thought to have arrived on a ship from the South Seas, though nobody could be sure. Soon, the royal menageries from both the Tower and Windsor Castle were sent to the new zoo. Although they were for scientific study, such places operated much like living museums, as moving versions of the taxidermy specimens found in natural history museums. Little about them was authentic: animals were often kept indoors lest they succumb to the cold, and they were never provided with their true natural diets – who really wants to see lions collectively hunt down a young antelope and tear it to pieces? While some of the draw of the Tower menagerie was the violent wildness, people paying to feed a stray to the lions, attitudes shifted as a burgeoning sense of animals as sentient beings developed. Still, as more zoos opened across the world, an important part of their provision was that they were open to the public as places of education, however light.[19]

Where the advent of the railways had given people from the cities access to far-flung parts of Britain, the development of the car and its mass ownership afforded a new way of encountering wild animals that was a step beyond: the safari park. This simulated the safari of Africa within the enclosure. The word 'safari' seems to have been introduced into the English language only in 1860, when it was used by Richard Burton – the explorer rather than the twentieth-century actor – in his writing on lakes in Central Africa. Burton was a polyglot who was fluent in several languages by adulthood, and during his travels he translated many texts, including the first *Kama Sutra* in English. 'Safari' came from the Arabic *safar*, meaning 'a journey', which became *safari* in Swahili. Burton himself belonged to an emerging school of people who partook in exploration for the sake of adventure, many of whom had been in the military and used skills they had picked up in navigation, survival and first aid to explore the most exotic and remote of locations. This, during the imperial reign when Britain had a grip on large parts of the world and imported all sorts of new tastes and traditions, charged a whole genre of fiction featuring swashbuckling tales of surviving in a dangerous place, among them *King Solomon's Mines* and *Tarzan*. Those who had the means and the connections could go on safari in what was then British East Africa, a place with a reputation for being wild and untamed enough that to survive it was a measure of masculinity. Big-game shooting was a popular pursuit for those stationed in this frontierland, feeding a global commercial trade in ivory. But by the early twentieth century, interest back home had been piqued and big-game tourism had taken off, spawning not just fictional adventures but travelogues that further fetishised the exotics of empire.[20]

In the decades that followed the Second World War, new opportunities arose at home, those that offered something more immersive than just observation. The wars had left their mark on the countryside – land and buildings had been requisitioned to contribute to the effort to produce enough food and materials. In the aftermath of the First World War, with the government raising funds via increased inheritance tax, and with compounding issues such as loss of heirs through death in service, the aristocracy found themselves in many cases unable to manage land in the way they had previously. Country houses were destroyed, and over the years more were handed over to the state and ended up with organisations such as the National Trust. The old parks that had been maintained at great expense in the eighteenth and nineteenth centuries had to start to pay for themselves, and their organisation and space – extensive areas of grass interspersed with watering holes and thickets of trees – made them ideal candidates as new open-air zoos.

In 1966 Longleat in Wiltshire became the first safari park to offer a wild drive-through experience, and it is still going today. It offers the opportunity to see a variety of animals grazing on the rolling hills of a Capability Brown landscape that would have been a sheepwalk back in the day. Much of its landscape park is preserved, still surrounding the impressive Elizabethan country house that was constructed for Sir John Thynne. The idea for a safari park didn't actually come from Thynne's descendants but from the owner of a circus, Jimmy Chipperfield. Chipperfield was from a long line of menagerie and circus owners: his ancestor, James Chipperfield, had been performing with a menagerie of animals as early as the Great Frost of 1683–4, when the Thames froze over and was celebrated with a carnival. Chipperfield had already set up zoos, and in agreeing to help set one up in Uganda had been provided with a licence to catch live animals in the national game parks there, in lieu of payment. Over the course of his expeditions, he came to the conclusion that what was wrong with the traditional zoo was the way in which the animals were displayed, that the cages and glass made the whole thing rather two-dimensional. Chipperfield envisaged an experience where people could drive through an enclosure of lions, stopping as and when they wanted for photographs, and seeing them in as close to their natural 'free state' as possible. It was not a hard sell to the aristocracy, some of whom had already contacted Chipperfield and implored him to help them create animal attractions on their land as commercial enterprises. In the end, Thynne, the then Lord Bath, was both amenable to the prospect and had suitable land, and the two hatched a plan to have fifty lions loose in the old park of Longleat.

The biggest barrier, at first, was procuring enough lions. Chipperfield attempted to import them not from the wild, but from captive populations across Europe, without actually telling anyone what they were for. He had great difficulty, until word got out. There was uproar, with people believing the lions would escape and attack all the sheep in the hills, or that rats would surely quickly overrun the park because there are no vultures – natural scavengers to clean up a carcass – wild in Wiltshire. In the end, the lions of Longleat came from captive populations in Germany, Spain and Israel, as well as ten who had been used in the 1966 film *Born Free*. None were wild. In time, other species and areas were added to the experience, and Chipperfield is the progenitor of most of the 'safari park' zoos in operation in the UK today.

Saving the wild

Although best known for the spectacle of seeing creatures from afar, zoos and safari parks have played an important role in the conservation of animals in other places. As they proliferated and became more organised, they were recognised as important repositories, animal seed banks if you will, of species that were disappearing through hunting, habitat loss and sometimes because no-one was looking. It is difficult to know how many species we've lost that we were, collectively, completely oblivious to. We wouldn't necessarily expect to find fossil evidence, or archaeological evidence if it was not an animal that played a part in the way we lived. In this respect the zoo plays a necessary role in the conservation and preservation of life. By the latter half of the twentieth century, thoughts were turning to the conservation of the species we had on display, and networks of breeding programmes sprang up that have been instrumental in saving some from the brink of extinction, though some of these are not always quite what they seem.

One animal that owes its survival to captive breeding is the stocky, beige and quite punk-looking Przewalski's horse. These were 'discovered' roaming the Mongolian plains in the 1870s and are proclaimed to be the last of the wild horses in existence, distinct from donkeys and zebras, which belong to a separate subspecies. They were found by a Russian explorer, Nikolai Przhevalsky, and first described in a Russian journal by J. S. Poliatow in 1881. The journal *Nature* followed this up in 1884 with an English-language article that described them as an 'intermediate between the true horse and the asses', stating that they were found in only 'the wildest parts of the desert'. At this time there was just a single museum specimen in collections, in St Petersburg.[21]

Efforts to trace the domestication of the horse have centred on Central Asia, particularly on a collection of horse remains found at Botai in north-central Kazakhstan, where it is known that 5,500 years ago horses were being harnessed, milked and corralled – it is the earliest evidence for horse domestication in the world. Researchers expected that sequencing the genomes of the Botai horses would reveal the genetic ancestry of all modern breeds that must have originated there and then diffused throughout the world. But they got a surprise: the Botai horses clustered not with modern horses at all, but with the Przewalski's horse, including both modern and nineteenth-century museum specimens. What this suggests is that the Botai horses were a domesticated horse that eventually went feral and was not used again by humans and *became* the Przewalski's horse. When they looked further, they found hints that the origin of all other modern horses lay not in Mongolia but in Hungary, some 1,000 years after the Botai horses seem to have been loosed back into the wider landscape.[22] The story doesn't end there, though, because just eighty years after it was discovered, the Przewalski's horse was extinct outside of captivity. What was brought back and released into the plains – for at least the second time in its history and with the helping hand of the ZSL – is genetically increasingly inbred and has seen the introgression of modern domestic horse genes at a rate of sometimes over 30 per cent.[23] So, while it has been 'saved', it's not quite the story of a wild and pure marvel discovered roaming in an untouched refugia that has been nurtured back to healthy numbers. It's more like a repainting of the *Mona Lisa* – all the looks but something's missing. We've unleashed a facsimile, a new edition. An animal of a lineage interrupted down to its very bones by human tinkering.

Yet, does this make it any less worth saving? It may not be a pure wild, but by now I think we know that the idea of that doesn't really exist anywhere. The Przewalski's horse is part of our story as much as the story of the plains where it was found. By studying them we have found a hidden part of the story of how we became close to horses: that we harnessed them before giving them up for a while. Rather than disappointment that it is not a survivor of times before, it should inspire awe for revealing how we lived with animals in the past, illustrating that things are never linear and that our world is built on ecologies of which we are an integral part, whether as progenitors or disruptors.

Feral

The Przewalski's horse is not the only animal that has journeyed from wild to domestic to feral, though its companions have different reputations. The horse is seen as a wild wonder, a link back to a time when such megafauna roamed great plains unhindered. But the pigeon, another leftover from our cultural moulding of animals that has taken its own path outside of human mediation, is the rat of the sky. All pigeons, and there are hundreds of domestic species, come from the rock dove. There really is no difference between a dove and a pigeon, except a pretty name and a better reputation. The difference between the horse and the pigeon is one of proximity – pigeons are now largely dependent on human waste and so are found in huge numbers in cities. Racing pigeons came into being in the nineteenth century, and their frequent escape into the cities in which they were kept is what has led to their proliferation. The rock dove readily nests, as the name suggests, on ledges, and the stone walls of the city are worthy substitutes for a species that doesn't need a nature that is green. The pigeon leaves its own waste everywhere, is generally *invasive* and is now seen as a hygiene issue. The first anti-bird-nesting spikes, those vicious-looking things adorning suitable perching spots on buildings, were patented in 1949 and intended to stop pigeons and corvids from getting too comfortable. Although they are now mass-produced and an incredibly common sight, these measures are not particularly effective. For one thing, there have been quite a few instances of birds ripping off the spikes and using them to build nests, as the spikes mimic those of hawthorn, sea buckthorn and blackthorn that birds such as magpies nest within for protection.[24] This would be surprising, were it not for the fact that it is now estimated that there is more human-made material in the world than there is natural, and large populations of city-dwelling birds who live on our waste continue to increase in number and readily use the available unnatural home-building materials.

The advent of bird spikes, the banning of feeding pigeons, the boxes full of poison placed along the edges of walls – all are indicative of a trend that emerged in a sort of parallel or mirror image of the conservation movement that sought to protect nature. Human spaces were only for humans and the very few chosen animals that had the privilege of walking among us – these were generally pets, and certainly not rats, pigeons or insect pests. The reason this shift occurred in the mid-twentieth century has got to be related to the way in which technological advancement meant we replaced so many animals with non-animal counterparts. Whereas the city in centuries before had been alive with animals that were traded at markets, horses pulling carts, dogs turning spits, pigeons carrying messages, they were all replaced, by cars and bikes, private kitchens and Royal Mail.

Severing ties

At the beginning of this period the wild was presented as something that was free from people, green and often rugged, and could be looked upon as if it were a painting. The places people thought fulfilled the ideal of this were often those that had been altered in the past, deep or otherwise, but in the face of increasing urbanisation people bought into the idea that they could go and experience the beauty and peace of the wild in the countryside. It also became a health issue, leading to the creation of municipal parks in urban environments, which gave people an approximation of nature for their leisure time; outdoor pursuits rose in popularity. But increasingly there was a disconnect between us and what actually lived in nature, the birds and the beasts. Their removal from our daily lives led us to seek to remove the hangers-on from areas that were now supposed to be cleansed of the moral and actual filth that accompanied particular animals. This wasn't confined to the city, either. Out in the countryside, animals were under threat from many angles. Those that damaged crops, like rabbits, fell victim to deliberate control measures that, excuse the pun, ran wild and wiped out millions. Myxomatosis was introduced in France to control the rabbit population, before it leapt the Channel and decimated the numbers here in the cruellest way. Others like otters fell victim to pesticides intended to control insects.

What happened was a final severing of us from the wild – we simply no longer understood what it looked like. It had come to be represented by a complicated mass of ideals – it was green, it was beautiful, it could be dangerous, it shouldn't invade the human world. All the while, conservation efforts meant we understood we'd had a hand in the loss of both animal and plant life, and we sought to find vestiges of the old world. A shame developed, one that exacerbated the idea that the wild was something better off without us and stopped us from considering the myriad ecologies that already only existed because we are and always have been part of the natural world. We had undeniably done extensive damage, but this attitude seems to have prevented us from connecting the dots to understand that just because it's different doesn't mean it's not worth protecting. So, we grubbed out orchards and sprayed the verges, but other more incidental habitats like railway sidings and abandoned barns became places where wildness was out of sight and out of mind. Just like the medieval waste before them, these spaces became thought of just as holding places until something useful could be done with the land, only instead of human commoners it was animals and plants that were moved off without thought when development was undertaken to 'improve' them. Now, the road verge is made beautiful with a carefully curated mix of 'wildflower' seeds, though it was probably perfectly serviceable before – it was just a bit rough around the edges with coarse grasses, meadowsweet and 'boring' plants like cow parsley. For all our hand-wringing about the state of the world, much of what followed was a papering over of the cracks and kicking the actual problem a little further down the line – making something look lovely without thinking about what's best. We'd ended up in a world where everything had to have a use, a place, a reason, even if that was 'to be wild'. We stopped understanding that our whole landscape was already curated and that the best way forward would to be to work with it rather than wish for a time when it wasn't.

And so, we arrived in the twenty-first century with a divided legacy: a land where barriers were entrenched, fiercely defended with signs and barbed wire, but also with perceptions of who should or should not be in that rural space. We have no fear of wild predators: we told stories about them for so long that they became fictional, and we don't believe there is any danger there *even* if there might be. We have become the predator. This is the kind of rhetoric that swirls around issues regarding our attitudes to a wild Britain in the present; this is where it has all led. Shame, longing, curiosity and entitlement form the cornerstones of our engagement with it. Shame because we are able to recognise that we've had a massive impact on what lives in Britain and why; a longing for what we believe to have disappeared; curiosity because we no longer feel immersed in a world we think we should naturally understand; and entitlement because we feel that these things should belong to us all. Of course, each of these issues is far more complex than that, and all are related.

Our shame means we are often dismissive of peopled places, whether or not they might actually have ecological value that we're ignoring. Our longing blinds us to reality and instead replaces it with a rose-tinted view of something we think was here in the not-too-distant past, without understanding it has always been in flux.

Curiosity makes the exotic more attractive than the realistic, contributing to our longing for a set of fauna, in particular, that were once abundant but would now be out of place, and meaning we seek out and highly value anything we perceive of as close to a world without human impact. And entitlement feeds a belief that not only should we be able to get back what we think we've lost, but that we should also be able to see it with our own eyes, which sits uneasily alongside our shame in the places where people are and have been. We need to reconnect with the processes that brought us here, to make informed decisions about where we might go in the future.

7

THE NEW WILD

The wild of now

During the global pandemic that dominated our lives beginning in 2020, newspapers turned to the natural world for some lighter news, and a raft of articles celebrated the return of the wild to the cities. While the world was confined indoors, animals like wild boar and deer became bolder, setting foot inside spaces where they are not normally seen. Manicured green spaces were left to feralise while the people who normally tended them were made to stay at home. But there's a reason we don't normally see deer in cities or scrubby parks high with grass and brambles. Cities are designed for people, and although animals do live in them, in the streets and the trees and the sewers, they are not considered to be the 'right sort'. For thousands of years we've been trying to keep what's ours *in*, and what's wild *out* of, our space. Although bears in Italian cities and dolphins in Venice were nice to read about, everyone knew it wouldn't last. Among the many things highlighted by the pandemic was our strange relationship with the natural and wild worlds, and all the inherent contradictions and entanglements. We have consistently sought to encounter animals on our own terms, expanding into new space but expecting them not to expand into ours. Large animals that were hunted for food were often to be found in clearings that we manipulated to ensure a good supply. Later, this led to the management of forests and the legal enclosure of land. Gardens grew from a desire to create controlled spaces for quiet contemplation as well as a reliable source of plant food, safe behind walls and fences that kept other animals from stealing. As cities developed, people and animals were employed to curb the attacks of pests. Outside the cities, animals were preserved first for hunting and human use, and much later for their own sakes in designated reserves. In the twenty-first century these measures have resulted in a specific zoning of spaces where particular bits of wild are and are not allowed. This has become our normality: we cover buildings with pigeon-proof spikes and fumigate wasp nests, but travel to safari the countryside in the hope of glimpsing some 'real' wild.

This is the last stop on our journey, the final story of where the wild went and where it might be found. We know that the wild we have now has adapted to and been shaped by us over time, that there is very little about the way the jigsaw is stuck together that would be recognisable to the craftsperson of Star Carr. Much is made now of how we can try and get back some of that past, in the hope that it is a solution for the future. What follows is a look at where the last 12,000 years has led, and at some of the solutions now being proposed as well as new threats we face from the wild and its kin. Our nostalgia for the past means that for many the ideal would be turning back the clock to a point in time we like the sound of. Yet whereas people in the past had their own challenges, and we know that the climate certainly has its own variations, there is no precedent for what comes next. We are part of the problem but we are also embedded within the possible solutions.

Rewilding

In more recent years we have collectively begun to recognise the scale of the impact we've had on the world beyond us, in the form of not just a changing climate but also the rate at which we have lost plant and animal species, once we stopped simply categorising things and started to track their prevalence and behaviours. We have begun to rediscover our place in the order of things and have developed an understanding that long ago, we lived in and as part of the world. As time wore on and this was lost, the environment surrounding us became a backdrop to our lives, where we observed change as a largely localised and short-lived phenomenon. We had become so used to the environment bending to our will that we had lost sight of the value in letting it be. But as we now grapple with the realisation that we *need* natural processes to survive, and most of us at least accept that this must be integrated into how we shape the future, we have looked back to see how things used to be and might be again. One approach to emerge has captured the public's imagination: rewilding.

Rewilding as a concept is really only a few decades old, conceived in the 1990s in North America and intended as a way to tackle fractured and damaged ecosystems by joining up landscapes and introducing more large predators to balance out species and reinvigorate natural processes. Since then, it has expanded conceptually and geographically and since the early noughties has been gathering pace in Britain. The European perspective tended to focus not on predators but on the reintroduction of large herbivores that have historical and sometimes extinct equivalents. It was hoped that these could provoke the type of succession described by Franz Vera: that the old wildwood would have been naturally punctuated with open spaces that were kept that way by the constant grazing, trampling and browsing by large mammals. Herbivores that eat woody plants are also useful in a warming world as the clearings help to mitigate the higher likelihood of forest fires breaking out. In places like the Iberian Peninsula, where forest fires are increasingly frequent, this could be incredibly helpful. The drivers behind an increase in forest fires in that area are not solely down to a changing climate, although it undoubtedly compounds the issue. In rural Portugal, for instance, the mid–late twentieth century saw a surge in replacing native forests with eucalyptus plantations to satisfy the growing wood pulp industry. So much was planted that it is now the predominant tree species in the country. After these were established, rural areas experienced depopulation as people left to find work abroad or in cities, leaving plantations unmanaged, with deadwood that catches fire and burns easily, especially as the trees cultivate a soil that is dense but very dry.

For Portugal, then, 'rewilding' in the form of replanting native species and, in the absence of people, 'management' by large herbivores could help to mitigate for the rising temperatures that make fires more likely, although this would need to be balanced against the fact that eucalyptus is central to local economies. We have very few forest fires in Britain (although they are on the rise), and a much more densely populated landscape. We also, especially since the pandemic, increasingly want to live in the countryside and work remotely. What does the concept of rewilding look like around us?

The rewilding buzz in the media has done much to heighten public recognition of the realities of our landscape, but it has also become a bit of a catch-all term for any type of remedial habitat work, and for many it just means leaving nature to it. Among practitioners, debates rage about the form that rewilding should take. For some, it is the removal of all traces of us, a measure that would allow nature to reclaim an area and find its level after enough keystone species – those that play a critical role in maintaining the structure of an ecosystem – find their way back. It is this that has captured imaginations, because it lends itself very neatly to how we have come to understand what the wild is. Critics of the approach believe it serves only to reinforce the nature–culture/human–environment divide, placing people on a plane of their own with no role as part of nature or the environment. I am inclined to agree. It doesn't seem realistic for the landscape of Britain, already so utterly shaped over thousands of years by people, nor does it seem possible to achieve. If we were to follow it to its logical conclusion, we'd end up depopulating the entire countryside and living fully urbanised lives, which would serve only to completely sever us from the nature we so desperately want to feel close to. We also know that soon after the last Ice Age ended 12,000 years ago, long before farming made its way here, we were actively creating and managing these clearings ourselves, working alongside the large elk and aurochs. At the other end of the scale are approaches that accept rewilding as wholly within human control. In this model we can make overtures to wilderness but have to compromise on how peopleless it really is. For other approaches it's somewhere in the middle, with rewilding something that has to work for both humans and the environment in equal measure – the balance between retaining eucalyptus as part of the local economy and removing it as mitigation for forest fires, for instance. Another example would be the reintroduction of predatory carnivores possibly harming livestock, but rural populations affected by this may be encouraged to adapt to a new stocking regime which allows both to co-exist.[1] There is no either/or in these scenarios, just a better way of doing things.

The downside of having become the solution *du jour* is that rewilding projects, whatever their character, are under enormous pressure – there is a huge weight of expectation on projects to deliver answers, to fix things, but also to give us opportunities to see things we won't see in other places. We want them free of obvious human intrusion, but they should also look nice. Nature should recover but largely be filled with unusual animals we wouldn't otherwise see – 'ancient' forms of horses and cattle, beavers or perhaps even wolves. In a sense, these projects are the new zoo. Contrary to the way they are often portrayed in the media, most projects don't claim to turn back the clock to a time before human intervention, nor do they attempt to pinpoint a moment in time to replicate. This would be a mistake: how do we decide which time was 'best' and how could we possibly recreate a world without humans in a world that is made by humans down to the very soil chemistry? There must be a way for the world to be wild*er* without necessarily striving for a pure wild that rejects everything that we are as human, and everything we have been in the way we have lived.

The total removal of people from landscapes also presents something of an issue, because these areas, however wild we try to make them, *do* need to make money in order either to replace funds that are lost through the cessation of agriculture or because setting the wheels in motion to really make change is expensive. There are various funding models, each with its own characteristics and issues. One is by charging to go on 'safari', as at the Knepp Wildland project in West Sussex, to *see* the wild in action, perhaps even camp in it. That people feel entitled to see it but require a sense of solitude to *believe* in it means that some parts of rewilding projects become designated for some people and not others, with those who pay having access to the best or 'wildest' experience. Like medieval and even Roman landscapes, it privileges some people over others and continues to elevate the wild to the stuff of myth and legend.

The eulogising of rewilding, the wildspeak, is perhaps its most fundamental issue: the term 'wild' is *so* loaded, particularly with the sentiment that humans are the enemy, that in a landscape that has been wholly and unrelentingly shaped by people is it really a useful term at all? By continuing to privilege what we feel is the most wild over everything else, aren't we perpetuating the very attitude that continues to separate us from the natural world and has led us to crisis point? If these are to remain living landscapes, they are going to be affected by humans whether or not people safari in them or not, or whether there are old buildings and ruins, or whether there are farmed domesticates. What we have now is the product of thousands of years of interaction and entanglement between the environment and everything that lives in it. There are things here that have made the islands their home and have even come to take on core cultural roles, animals that have moved into our old spaces, and plants that have used our modification of soil to their advantage. The way animals and people move around, inhabit and attempt to control territory has led to the design of tools and even plants to use as barriers, which have in turn spawned their own unique ecosystems – linear hedgerows or particular communities of plants that occupy the understory of coppiced woodland. To use a term that so heavily implies that things are better off without us wipes away not only *our* story, but the story of the wild that is, was and will be. Rebalancing our relationship with the environment means reintegrating rather than removing ourselves, and finding a way to make us more resilient to the future rather than fit for the world of 6,000 years ago.

Producing wilderness

The success of the idea of rewilding as a solution for both the planet and our own health is in no small part down to the seismic shift in our way of life brought about by the COVID-19 pandemic. After a period of weeks of enforced domesticity in which we were ordered to stay at home, people began to yearn for outdoor space and some kind of reconnection with nature more than ever, while rules on social distancing meant this needed to be away from the crowds, resulting in an exaggerated effort to find the wild. And because beauty had long been the face of wild things, it led to everyone heading, ironically, to the same places. Huge numbers of cars took over the verges of the approaches to well-known bits of the national parks. Honeypot sites were overrun with people because for years we'd promoted them as tourist destinations of an ecological value over and above other places.

In these promotional narratives the wild had become something to be packaged and sold, its character emphasised by repeated use of the words 'rugged' and 'remote'. The levels of 'purity' and 'naturalness' became measures of how worthy a landscape is of attention and protection, but it was almost without regard to the *actual* naturalness, only to what was perceived. The word 'wild' makes an area appeal in terms of what people think they want to see – proof that nature can take care of itself, to alleviate our guilt at the hand we had in damaging it. Appending the word 'wild' onto things does also mean that it must conform to our aesthetic expectations as well. After all, there is little market for leading nature safaris into abandoned supermarket car parks, even if sometimes the quirks of its history mean it harbours rare solitary bees or perhaps even ground-nesting skylarks. Sometimes visitors even expect the rewilded areas to have a kind of neatness about them because they are so unlike the nature we are used to, all scrubby and apparently borderless and not really reflecting what wildspeak would have us imagine. In some places the animals themselves are the draw but aren't necessarily all that they seem. The herd of white cattle at Chillingham Castle, for instance, are a visitor attraction often spoken about as the last living descendants from the aurochs that once roamed Britain, imbued with an ancient wildness. Genetics, however, tells a slightly different story. Although there are some so-called 'primitive' markers in their DNA, we know that the aurochs indigenous to Britain were hunted out by the Bronze Age, and that cattle here were domesticated from elsewhere. The genes of the white cattle of Chillingham cluster closely with other domestic breeds,[2] so the cows are essentially a herd that's been around for a very long time, a bit like Przewalski's horse. This doesn't need to be and shouldn't be disappointing, because I'd argue that for both breeds the *cultural* story, the one that reveals the interaction between our species over millennia, is far more interesting than any claim to being the last of the wild. The reality is better than the fiction, but for so long we have been burdened by the notion that there is no wonder to be found in our journey.

The branding of the Wild Atlantic Way of the west and south-west coast of Ireland is another illustration of this in action. And this is despite the fact that the term 'wild' has a rather darker history in Ireland, rather than being an affectionate term for aesthetic beauty or rich flora and fauna. It was the language of colonial rule, the perpetrators of which saw a culture of barbarism that needed refining. It may have morphed over time to become imbued in the land, its mountains and bogs, but is a term that is ultimately rooted in persistent violence and denigration. The term's history is at odds with its contemporary marketing, but is evidence of the way in which the word and idea has been elevated. Where once it was a term that made being outside of the human world bad, now it is seized on to exemplify the good.

The Wild Atlantic Way now brings in significant numbers of tourists, having largely sprung from nothing not much more than a decade ago. It is the world's longest coastal driving route, running for more than 2,500km from Donegal in the north, down the west coast and around the south coast to Cork. The route was formerly simply a collection of small country roads that were linked all the way up the coast, little known outside of their local areas. But here, the word 'wild' has had such power that the route was traversed by an estimated 8.4 million people in 2019, 3.4 million of whom were overseas visitors, the majority of them visiting Ireland for the first time.

Conveniently situated along the interface between land and sea, the route is easily both conceptually and geographically bounded, which helps to make it readily understood as a physical entity. As the full force of the Atlantic Ocean swells against one of the most westerly seaboards in northern Europe, this network of roads allows visitors to get up close and personal with some of Ireland's most spectacular countryside, land and water. The northern end features a long-standing attraction, the columns known as the Giant's Causeway, formed tens of millions of years ago by molten basalt that seeped up through cracks and eventually cooled into geometric patterns. The bits between hotspots like this are no less breathtaking, home to puffins and basking sharks, with long beaches of white sand and high tussock-covered cliffs. Despite this, it has taken marketing the whole coast as 'wild' to draw people in.

'Wild' Nephin

There are, of course, also rewilding efforts along the way that are making gains for nature. On the western side of Ireland is Nephin Beg, a range of low mountains and internationally rare blanket bog that was designated as a national park in 1998. It already featured several species that are legally protected under EU legislation, among them golden plover and shining sickle moss, named for its hook-like fronds. Fast-forward nearly twenty years to 2017, and the park was expanded to take in thousands of acres of coniferous forest controlled by the government's forestry operations, and the following year the park was renamed Wild Nephin. It was explicitly envisaged and advertised as the 'one true wilderness' of Ireland, where non-interventionist management strategies would be employed to 'rewild' the land. The Irish naturalist Robert Lloyd Praeger is often quoted on the topic of Nephin Beg: writing in the 1930s, he believed it to be one of the most desolate places in all of Ireland, but found that thanks to 'the mystery and majesty of nature' it was 'not lonely or depressing but inspiriting'.[3] It very much echoes the Wordsworthian pastoral – a rugged and isolated place with the capacity to override human influence in a divine way. But while it has a history of inspiring wildspeak, like all areas in these islands it bears the deeper marks of people over time.

After Praeger's words were written, the land was exploited as a commercial forest from the mid-twentieth century. Those coniferous forests were not the ancient pines of yore but regular and organised stands of imported Sitka spruce and lodgepole pine. The creation of those plantations didn't just involve planting trees, but in a place like Nephin Beg required extensive intervention to make the soil suitable for maximum growth. Drainage schemes were created with ditches, and chemicals leached into the land that attempted to make the acidic blanket bog more nutritious for saplings. In 2002, four years after the designation of the area as a national park, the European Court of Justice was heavily critical of Ireland for the overstocking of sheep here that, since the 1980s at least, had contributed to the degradation of habitats. Soil had been worn to bare rock; heather, an important nesting environment, was disappearing; watercourses were silting up and becoming contaminated. The response was a positive step: reducing sheep numbers and now, of course, an intention to let nature wipe away the effects of commercial forestry.

Not many people would be sad to see the homogenised and orderly, video-game-like trees of those forests removed in favour of something a bit more organic, but it is possible that in pursuit of a naturalness above all else, other valuable evidence of the past may be lost. There is archaeology here dating back into prehistory: the records we have of this are reasonably sparse, but that's because there has been little systematic surveying rather than because people were never there. We generally value archaeology and heritage and put protective measures in place when we know they're there, but equally, we know that past human activity has the capacity to affect the ecological character of a place for thousands of years. What does and does not grow is often symptomatic of what lies beneath and beyond, driven by things we've done to the land in the distant or recent past. What went before has the potential to privilege invasives that stifle many of the plants that we want to thrive in a rewilded world. In short, it's entirely possible for the *wrong* nature to heal. An area that is *truly* left to its own devices will grow and morph until it finds its level, but it might not be what we want it to be. It is clearly not the case that lots of rewilding projects are filled with terrible plants and animals that do more harm than good, but it's also not the case that every abandoned industrial site needs saving and redeveloping. It's a much more nuanced and complex picture, and each project requires its own approach.

The pervasive idea of how nature heals, the feeling that it has the blanket capacity to eventually triumph over us, is clouded by places like the Amazon rainforest where cultures thrived before disappearing back into the jungle, only to be discovered in more recent years as blocks of stone visible on aerial scans. But the reason that those places have been so successfully subsumed back into nature is because the intensity of landscape use and the population density around them were incomparably small when compared to our collection of small and exhaustively exploited islands. If the British Isles were abandoned there would be plants left over from agricultural histories that wouldn't be immediately swamped: smatterings of oilseed rape, oats or wheat; lots of nettles and brambles; and invasives like knotweed. Perhaps these plants would then render the area in some minds as somewhere that needed to be wiped away and replanted from scratch, but they would provide vital habitats for those who require that niche. There is the lingering threat that some species would overwhelm others and become dominant, at the expense of those that would support a more diverse ecological network. At Nephin, then, those who manage it employ both cutting and the herbicidal control of plants, like rhododendron, that would otherwise damage the peat and prevent favoured habitats from forming or recovering. It is a responsible way to make sure that Wild Nephin achieves the gains for nature that is its primary reason for being, while also conforming to various directives that stipulate land should be managed to fulfil the needs and niches of particular birds and plants. But we can't really consider this to be 'non-intervention', surely. Actual non-intervention would result in a completely different character. Nephin required a gentle steer to help nature find the level that is right for the area's particular circumstances.

Credit where it's due

A major consideration with ecological restoration, rewilding and similar projects is that it all needs to be paid for in some way, and it is expensive. The vast majority of projects aren't about just leaving things alone to see what happens. It is for this reason that some projects offer certain activities, such as wildlife safaris that come with a fee, or try to find sponsorship, which adds to the pressure they face to please everyone. A great deal of natural heritage in this country is funded by bidding for money available as part of the National Lottery, of which a key requirement of projects is that they involve people in some way. Government funding is sometimes available, but normally tied to time-limited aims. All of these schemes are extremely competitive, and only a fraction of the projects put forward are ever successful. In recent years, some organisations have begun to turn to carbon credits or offsetting as the source of their viability. The premise of these schemes, which have been around for a while, is that companies can buy credits, which then gives them the right to emit a certain amount of CO_2 or other harmful gases into the atmosphere. Caps on emissions mean that some top polluters, companies like Shell for instance, are obligated to buy credits to offset the harm they are doing to the environment in the course of their daily business. While most deal with becoming carbon neutral, steps have now been taken to introduce statutory *biodiversity* credits. These are purchased from the government at set rates, each one vesting in its owner the right to destroy a certain amount of a particular type of habitat, whether wood, grassland or moor. Credits are available to developers when they are unable to make a 10 per cent net gain in biodiversity either at their development site or on other identified land. So far, so sensible – it seems a straightforward way of raising the funds necessary to enable projects to make changes for the environmental good, making more gains than losses. But issues arise when things are monetised in this way; there is already a burgeoning industry in the trade of carbon credits. Because the environment is finite, so too are the credits, which means they can be manipulated just like any other money market.

The cost of biodiversity credits depends on the 'distinctiveness' of the habitat that is going to be destroyed by building on the land. How distinctive the biodiversity is depends on a combination of its rarity and species richness. Under this scheme, at the moment a single biodiversity credit can be purchased for £42,000, allowing a company to destroy a habitat of 'low' distinctiveness like scrub, going up to a massive £650,000 for destroying a peat lake, aka bog, with all of its stored carbon. All hedgerows, which seems to include everything from nineteenth-century blackthorn to Bronze Age hedge-topped banks, command a price just above the minimum.[4] The goal is really to encourage developers to do good for the environment by carefully considering the green spaces affected by their work, and so credits are considered a last resort. There is also a cheaper option, which is for companies to create their 10 per cent gain in biodiversity on someone else's site. In this scenario, landowners submit to a central database a parcel of land that has good potential for habitat restoration and name a price for carrying out the work. However, the early signs for the effects of these endeavours, in England at least, have not been that promising. So far, 91 per cent of biodiversity gains have been made within the footprint of each development, which sounds great on paper. But, in fact, these have taken the form of new recreational grass areas, as well as hedges on the boundaries. More than a quarter of the gain was made up of recreational grass that had developed on the edges, sown as an incidental aside to greenify the aesthetic.[5] Quite aside from the fact that expanses of homogenous grass like these are about as ecologically useful as a golf course, these tiny areas form distinct and separated units that don't link up, and experience huge pressure from the people who live there – at the expense of whatever coherent whole was there before, whether that was an old meadow rich in herbage or a brownfield site of buddleia or butterflies. I'm not arguing here that building houses is the ultimate problem. We need places for people to live and we need to recognise that some sacrifices are both necessary and inevitable. Rather, the issue with these sorts of schemes is that environmental benefits are distilled into neatly packaged credits, abstracting what it means to do actual useful work that moves us towards a better way of harmonising our needs and desires with those of a wilder world. It is so very easy to turn a blind eye to the destruction of ecological communities when they have been given a number on a scale, all of their character and knotty webs of relationships reduced to a single line of text on a spreadsheet.

If we're going to make our developments better for nature, surely it would be more sensible to come up with new ways of creating habitats that are good for both, rather than turning them into commodities that can be retrofitted around the way we already do things, all to be ticked off on a form somewhere. Moreover, these credits are a trade in immediate loss for the promise of a better future, mediated and set down by monetary value, one that means that what we consider to be lower-quality habitats will be lost in place of higher-quality ones. If this was replicated everywhere, a staggering number of habitats could be wiped out on a promise and a prayer that a nicer one will be created somewhere else. Even where off-site gains are made, they rely on someone else maintaining them for thirty years, with no clear consequences if they don't, and no plan for what happens at the end of thirty years. There is a clear need to perform restoration and conservation on damaged ecosystems, and in some cases create it anew, but what we do now needs to have a shelf life beyond the next few decades and take into account all of the things that are already here.

Accidental wilds

The bit of Cornwall I lived and worked in, the Tamar Valley, was the bit that most people don't go to. It wasn't the Cornwall of seafaring, surfing or holiday parks, but a slower and mostly overlooked borderland on the river. Our house was down a rough, unmade track, surrounded by woodland and filled with wild garlic that would hang heavy in the spring air. The house was damp, an almost unavoidable consequence of living in a thick-walled cottage built of granite and cut into a steep-sided valley. In contrast to our previous neighbourhood in the centre of Bristol, it was the rural idyll in the middle of nature that many people dream of, but it was really a bit of a myth. The wood hadn't even existed less than a century before. The house was riddled with radon, a poisonous gas which seeped out of the bedrock below and pooled invisibly in the lower floor. Hidden in the woods was the legacy of centuries of human activity: there were open mine entrances, and next door an old quarry had been reused as a Victorian refuse dump. There were artificial pools, ruined buildings and overgrown packhorse trails. Despite these vestiges of industrial use, the woods were alive with jays and foxes, butterflies and wild clematis. The space had been claimed by animals and plants in the absence of humans, and probably not for the first time; that space had seen the waxing and waning of human and animal influence, some species moving in as others moved out. Places like this show us that rewilding is *everywhere* in some way, in every brownfield site that has been abandoned, every mine that is no longer used, along railway sidings and clinging to the submerged posts of the crumbling old piers. It is in the gaps between what we think of as useful or beautiful space. I'm not suggesting that it's all of equal 'value' and that gutter moss is the same as a pristine peat lake, but it's the reality of our landscape and some of it deserves more attention and recognition. We don't tend to see it because we have elevated the wild to a thing of mythical beauty rather than the complicated and sometimes ugly reality that it is.

This is not new. It has been happening for thousands of years, human infrastructure disintegrating slowly and new wilds making their homes in what's left. We have seen it in the stories above about the beavers that moved back into the burned settlement of the Bronze Age Must Farm, and the owls that made their home in the rafters of the abandoned Roman villa in Oxfordshire. Today, it is the sycamore that grew into an icon of Hadrian's Wall and it's the lichen on Stonehenge. Quite often places like the abandoned quarry are left as they are because they are no longer so crucial that it is worth the effort to use them in the same way. In other, often suburban, cases land is left until the right conditions come along and it can be sold. These unmanaged places are sometimes impossible to access, but in others they become areas of informal use, which seem somehow to work in much more balance than those places that have an overt reason for being. Footpaths begin to deviate from their set route, as they may have done in the past. Brambles quietly expand their reach from the hedges into the grass beyond, enabling feral hops to expand out with them, and sheltering sapling trees that attempt to colonise a new woodland. Filled with dandelions, ragwort, little climbing vetches and rough grass, bordered by overgrown willow, elder and dwarf elms, these places are largely unnoticed because they aren't designated as somewhere we should pay attention to. But life there quietly goes on, the spaces become mini-migration routes for animals and plants that are no longer disturbed by ploughs or heavy grazers, and while humans are present, we interfere less. That is, until these places are sold or traded off for credits, destined to be made into something more productive or beautiful, at which point their quiet new communities move on again.

These places that we ignore are sometimes the wildest things we've got now, whether or not they're brewing a set of relationships that is objectively good or bad for the future. Because they are *truly* unmanaged there is no expectation of them to deliver anything at all in the present and they are akin to rewilding without an agenda. Many of them are not the best versions for the future, true, but some are harbouring the kinds of ecological relationships we haven't thought worthy of attention. They're not championed as rewilded places because they don't fit the idealistic and aesthetic version of what our collective imagination thinks it means to be wild, and no-one really thinks twice about wiping them away. In the grand scheme of things, we do need to balance the needs of development and housing with the threats we face from biodiversity loss and climate change, and not every patch of forgotten ground has the same ecological value. But, because we have categorised everything and distilled it into its most basic form, we seem to struggle to understand the bits of the world that don't fit the models we've built in our minds. Just because so much of how we view the world has a deep history, reaching back to when we first managed other beings for our own gain, doesn't mean it is appropriate for the world we have now. We may have been on this journey for thousands of years but we've found ourselves on a road that is not leading us to where we want to go. We need to dismantle some of the old barriers and ideas so that we can reintegrate with the natural world, to see ourselves as once and always a part of it.

Wilds abroad

This is the story of how landscapes and attitudes have developed in the British Isles and where we go next. It is unique to our topographical and historical conditions. Other places around the world have different stories, but also present opportunities to learn something of the way they relate to the natural world, even as the globe becomes increasingly homogenised.

Halfway around the world, Japan has the concept of *Satochi-satoyama.* It denotes an area with a lived history, which has been shaped into a mosaic habitat by human interaction with the environment over time. The area encompasses not just the rice fields and coppiced wood but also virgin forest beyond, old-growth forest that has never been significantly disturbed or cleared by people, with the exception of the occasional temple. It is recognised as something that has been created and driven by people but, crucially, has been shaped in a sustainable and balanced way. In this scenario, humans are seen as a crucial part of the continued success of an ecosystem that has developed in something of a symbiosis. Humans are not seen as the problem but as part of a dynamic environment that can and does change but is responsive and adaptable. Consequently, the ecological value of something is not measured by the presence or absence of people and their trappings, but on its own terms. Humans as part of a landscape with a mutual reliance, developed over thousands of years and allowed to change. Here, we tend to write places off once they are deemed to be sullied. It means places become damaged but also that they are not recognised when they have potential or have developed their own unique biodiversity. Given that our landscape has, as we have seen, already developed way beyond the virgin forest stage, it is clear that the practice could not be entirely replicated here. Perhaps, though, if we were to take the concept of *Satochi-satoyama* and apply it to our land, we might end up finding the courage to accept our role – the good and the bad – and move into a new relationship with the wild in between the bits that we think represent what it should be.

One particular project a bit closer to home hints at something like this, and shows what is possible with the abandoned bits we often write off. Along the train line running south out of the German city of Munich is a relatively new development of flats and shops, constructed within a garden-like space. A series of boardwalks weave through and away, venturing into what appears to be an undeveloped patch of waste ground, almost like the development just hasn't been finished. It's not unusual to find half-built developments, especially since the economic downturn of the late 2000s, though it is more curious for them to be brought front and centre and provided with a boardwalk for sojourning through. Among the scrub sit old railway tracks and sleepers, and even the rusted frames of old railway buildings. Through rusted windows the old control infrastructure is visible, and sometimes even the fixed furniture that provided past employees somewhere to work. These are the leftovers of an old train depot, long since robbed of its job. Although it seems strange, this boardwalk is no accident, and the development is not unfinished. The relationship between the old depot buildings and the greenery and scrub that has been allowed to grow where ordinarily it would have been cleared and made beautiful is the whole point. The entire development was conceived to include an ecological park from the very beginning, one that uses the buildings and tracks to form ecological niches in which new wilds can live. The ballast beds of the former railway prevent colonisation by fast-growing plants and have provided a habitat for sand lizards, for example. Buildings increase the vertical and horizontal variation in habitats, as they cause less shade than trees, providing nesting sites for birds at ground level as well as on signal poles, ledges and roofs.

This is an example of humans living in and with the new wild that has colonised our old spaces. It is an urban project but no less worthy than a rural one, and moreover it is a small demonstration of how we can have wild things around us without making their presence either a spectacle or a problem. However, at a national or large landscape level, there are other things to think about. With some animals moving in quietly, there are others that we seek to reinstate, and the knock-on effects of this can be considerable.

Think of the animals

We have seen changing attitudes to animals in general over time, but sometimes specific animals have also taken on different meanings and identities. The rabbit, for instance, which was once a revered exotic and is now a pet or pest. Then there are the animals that have disappeared or nearly disappeared, and in an age where we try to right the perceived wrongs of the past, we look for candidates to bring back. We have already done this for some species, and thinking about these gives some insight into how other introductions might play out. A good example is the red kite, a very large bird of prey with a distinctive forked tail that fans out and swivels to adjust the bird's flight. Historically they were very common across Britain, and because they are scavengers were probably viewed as beneficial in towns and cities for their consumption of waste associated with both food as well as animal processing. But like many animals that came too close to us, they were eventually persecuted, particularly for their predation of the animals that were reared for consumption or sport. By the turn of the twentieth century red kites had been persecuted so much that they were known only in a small part of Wales, where they were a major draw for visitors to the area because they were that rare. It was from here that in 1990 two birds were taken to Oxfordshire, along with several pairs brought from a red kite stronghold in Spain.

Since then, they have done exceedingly well – so well in fact that some have now been exported *back* to Spain to bolster the numbers that are dwindling there because too often they feed on the carcasses of foxes that have been poisoned to protect lambs. The success of red kites in Britain is in part due to the availability of food. While we are tidier than we have been in the past in the way we dispose of food waste, increased road use means there is a ready supply of roadkill. Continuing releases of game birds also mean that there are both small birds and carcasses to feed on, and the result is that not everyone is happy about the renewed strength in kite numbers. They are seen by some as detrimental to stock for the same reason as in the past. In contrast, though, the attitude is now cloaked in a concern for local wildlife, with those who are anti the red kite pointing out that they can predate on young hares and other small animals, those who are often seen as more deserving of our care.

In fact, this kind of argument against kites is reminiscent of those in support of badger culling, as well as those advocating pheasant rearing – that keeping the numbers of badgers and kites down enhances life for other wild beings. Since their reintroduction red kites have been rare in cities, where once they might have been found effectively cleaning up the streets, but this seems to be changing because people are actively feeding them in their gardens. At the moment, the red kite is afforded status in the public's perception: that of a successful reintroduction of something we once persecuted into oblivion. Given they have become so common, though, and with evidence that they are now habitually seeking out places where they are deliberately provided with food,[6] might they once again fall out of favour? The red kite was considered a pest for hundreds of years before its near-disappearance saw it morph into a worthy bird regarded with reverence, but perhaps before long it'll slip back the other way.

Daily death

What both the badger and the kite have in common is that they are welcome wilds until they begin to get a little bit too close, beginning to infringe on us and our property. We want them to be there and thrive but it's in an almost abstract way, to live unseen in 'the wild out there', but not to affect our own livestock or convenience. We generally accept the deaths of what we consider to be pests, such as rats, mice, mink, pigeons, wasps, ants, moths, slugs and snails. So, too, are the deaths of domesticated food animals acceptable, because that is, after all, the reason why most of them are bred and kept, and it all happens conveniently away from the public eye. This is intentional, a deliberate attempt to conceal from daily life what occurs *every* day – the mass killing of animals for human consumption. The public abattoir was a nineteenth-century institution that emerged in France in a move to regulate something that was seen as morally dangerous because of the violence involved.[7] In London, until 1855, live animals were slaughtered at Smithfield Market, something which ceased because of a cholera outbreak. Then, the slaughtering was moved outside the city, so that the public only had to digest, as it were, the pastoral sight of animals in the fields or the finished product of the butcher's work, rendered 'food' by virtue of skinning, gutting and a few well-placed chops, and not the unpalatable act of the loss of life itself.

This is perhaps the reason why we actually seem to have reverted back to some kind of mash-up of older attitudes to eating wild animals. It is taboo for many people because wildlife isn't seen as food but as something to spend time observing or learning about. The wildlife we do eat, such as pheasant, partridge and deer, isn't always really 'wild' and has often been, and to some extent still is, reserved for the rich. People haven't consistently eaten wildlife. We first turned away from it with the coming of domesticated animals in later prehistory, before the upper classes enjoyed it once more in the Roman period. It was lost again in the early medieval period, but after the Conquest once more became more of a food associated with the rich. It was they who had come to control access to the wild. By the eighteenth and nineteenth centuries there was a thriving blackmarket trade in game, but this dwindled and, by the turn of the twenty-first, we consumed very little. The only wild animal that is eaten by a large cross-section of society with any frequency is probably fish, though even much of that is now farmed, with 'wild caught' normally commanding a higher price. Some wild meats are considered premium, such as venison, which has been a consistently high-status meat over time, and others are more of a novelty. At the farmer's market or the butcher's you might find some slivers of cured boar for extortionate amounts, accompanied by the words 'wild' or 'artisan'. There are some people who believe vehemently that if you are willing to eat meat at all, you should be killing it yourself, and there are others who are 'wegan' – eating only wild meat and fish. There are some compelling arguments for eating more unfarmed meat: there would be environmental benefits to rearing a smaller number of farmed animals, for instance. We don't have enough to sustain every meat eater, but we do have an increasing abundance of certain wild animals that are doing very well in this mosaic landscape we've created.

Zoonoses

There is another side to our proximity to animals and the variety of niches they fulfil for us, one that takes away some of the control we feel we have. Despite our conceptual separation, we might be physically closer to them than we have ever been. We farm tens of billions of animals around the world on an industrial scale, and keep enormous numbers and types of pets, from the mundane to the exotic. We visit zoos, go on safari, feed birds in our gardens and in the park, and comment on how early or late the migratory birds are in any given year. If you were to look at your surroundings, there are likely to be myriad things with animal-derived products in them: the polish on the floor, the wool in the carpet, crushed beetles made into dye for cosmetics and food. Musical instruments are still frequently made with gut strings and use horsehair bows. Some televisions even have animal product in their LCD screens. We are sometimes surprised by these animal inclusions but tend to be comfortable with them. However, things change when the consequence of this intense relationship shows itself in a more visceral and unrelenting way.

The twenty-first century has seen a surge in the number of people living on vegetarian and vegan diets. For many there is an ethical basis and they simply do not agree with eating other living creatures, and for others the environmental footprint of industrial farming, mentioned above, is a critical factor. A few decades ago, though, there was another thing driving the swell in the ranks of the meat-free: fear.

In 1986, a new disease made itself known in cattle: bovine spongiform encephalopathy (BSE). It was found in a cow that had died two years earlier that had exhibited neurological symptoms including shaking, lack of co-ordination and aggression. By 1988, the disease had spread and the catalyst had been identified in the cannibalistic diet of bonemeal and meat that had been provided to them. There were rumblings that people might need to be careful about what meat they ate, but even as more and more cows succumbed to BSE, the government assured the public that beef was safe to eat. Only, it wasn't strictly true. Nine years after the first case in cattle came the first death from the human variant caused by eating infected meat, Creutzfeldt-Jakob disease (vCJD), in a nineteen-year-old man called Stephen Churchill.

This disease poses an insidious and silent threat. It is 100 per cent fatal and cannot be cured, but can lay dormant for decades before an infected person shows symptoms. Once the link was announced in 1996, there was mass panic about this so-called 'mad cow disease'. Sales of British beef plummeted at home and it was banned across the EU, the USA and other parts of the world. Public anxiety developed further when it was discovered the disease could be spread by infected blood, and several people died after contracting it through transfusions. There were sweeping changes in the management of beef for human consumption, culling, monitoring and bans on imports from countries with outbreaks of BSE. Public trust in beef products was shaken for years, and it was not until 2020 that the USA lifted the ban on importing British beef. Despite its gravity, the actual risk of contracting the disease is very low. To 2024 there were a total of 178 deaths from confirmed or probable vCJD in the UK, out of 4,856 cases referred for investigation, and 139 of these occurred between 1995 and 2003, with none confirmed since 2016.[8] There has actually been another human variant, simply named CJD, known since the 1920s, but there's no corresponding panic because it's not coming at us like an attack from the animal world.

The occurrence of vCJD was the latest in a long line of what are classed as zoonotic diseases, or zoonoses – those that are transmitted via animal sources. These include some famous ones like the bubonic plague, leprosy and rabies. The 1918 influenza pandemic, commonly known as 'Spanish flu', wasn't Spanish at all but probably from the USA, and sequencing has shown it leapt from birds into humans. It is the ancestor of all flu viruses known since, and also passed through pigs in the mid-twentieth century to become another pandemic-causing strain, H2N2. HIV is zoonotic in origin, as is Ebola, as are the respiratory viruses that have become known as 'bird flu' and 'swine flu'. Then there are zika, spread from infected mosquitoes and causing microcephaly in developing foetuses, and mpox, renamed from 'monkeypox', which has been around for decades but recently caught the headlines. We are especially sensitive to news of emerging zoonotic diseases in the wake of COVID-19. In the early days this disease was blamed on the seafood market in Wuhan, where the first cases were reported. It is now understood, however, that bats are a natural reservoir of this type of virus but that the data does not support a direct transmission into humans. Instead, it is thought that the virus emerged into the human population through an intermediate host or hosts, which is highly likely to include the pangolin, an unusual scaly anteater.[9] Pangolins are imported to China from South East Asia and traded, live, in markets such as the one in Wuhan, along with other animals. This brought them into close proximity to people, but also to other animals, through which diseases can pass. This is one of the problems with our nearness to so many animals: it brings species into close proximity that otherwise would never have met. Tourism, the global trade in exotic pets, animal parts and products, the disruption of habitats and forced removal of species, climate change and human development – all are pushing more animals out of some regions and into others. The way we live in this complex web we've constructed feels a little like experiments with a child's first chemistry set, mixing it all together to see what happens.

We are right to be concerned: our sensitivity to news of impending zoonotic diseases is well-founded. Approximately 60 per cent of all pathogens in humans originated in animals, and the source of 75 per cent of those that emerged in the last two decades before the turn of the twenty-first century lay in wildlife.[10] They've been around for thousands of years – brucellosis, the most common bacterial zoonosis in the world, seems to have originated in domesticated sheep in the Neolithic Near East.[11] And morbilliviruses, the type that cause measles, moved from cattle into humans in the sixth century BC, before moving from us to dogs, and from them to a large number of new and wild hosts.[12] While transmission between animals and humans is not new, it is happening more and more because, although we may not have anything left that we can call 'wilderness' in Britain, there are still some largely undisturbed habitats in other parts of the world, where animals act as reservoirs for pathogens that haven't had the chance to leap into humans yet. These pathogens can then spread between other animals or migrate with their animal hosts to other countries. Here, avian flu spreads annually through the wild bird population. You may have noticed that for a time in early 2023 'free-range' eggs were no longer actually free-range, owing to the housing order that was in place until April of that year to prevent wild birds from infecting poultry. While the risk to the UK's human population was and remains low, it is still possible that a flu epidemic will come to us from the birds we keep so readily as pets and for their eggs and meat. Leptospirosis, also known as Weil's disease, is a bacterial infection that can be spread by animals of all statuses and walks of life, from rats and mice to dogs and even perennial favourites hedgehogs and water voles. Even domesticated animals are reservoirs of disease, and the number of pathogens shared between us and them has been shown to depend not on how many uses we have for them or whether they live in our houses, but on how *long* it has been since they were first domesticated. In short, we share more diseases with cats and dogs than we do with something like pheasants. Domesticated animals are also likely to play a significant role in the transmission of pathogens and parasites between people and wild or feral animals.[13] They act as conduits for disease to move between hosts, in ways that we may not always understand.

It may seem like this means we should be further separating ourselves from the natural world, that perhaps it makes sense to have big spaces for animals and other spaces for us. It's just not practical, though, and isn't going to happen. Part of the issue is that we've already become so fixed in place, in our ideas as well as the way we have constructed the environment around us, that we've lost the will to adapt. Instead, we expect the environment to do the work for us, to heal itself while we carry on with whatever we were doing in 'our space'. Ultimately, the solution is not going to come from dropping in a couple of animal or plant species simply because they used to be there, and, thankfully, most projects don't do this at all, although illegal releases of popular species do happen. It's all a symptom of our disjointed thinking about nature and the wild. Rather than rewilding Britain, I think it is time to reweave ourselves through it, following the threads of our story to reintegrate humans as a working component of the systems that make our world so we can *all* face the responsibility we have to pivot towards a new future, not away from the shame of our past harms.

Using the past

I don't believe we should be trying to recreate the past as a way of making the future. I actually think relying on it so heavily is to the detriment of the good work people are trying to do, because it evokes such strong feelings about what is right or wrong, native or not, what should or shouldn't be allowed to be there. It can't work. We cannot go back to a time before people, because whatever we create now has to be able to function with our continued presence. And we can't choose a single point in time that we think of as ideal, because as we've seen, things have been in flux, moulded around the knotty ways all of the elements have interacted together – climate, species introduction and attitudes. Neither can we insist on things being 100 per cent 'authentic' and what would be here *naturally*. Our landscape became a kind of natural museum over millennia. This is not a situation we found ourselves in overnight.

Archaeology as a tool for informing our re-engagement with the wild isn't confined to showing us species that used to be there. It is frequently employed to prove that a place has had a long history of human occupation – archaeology is, after all, the study of the stuff people have interacted with and left behind and not of the natural world in and of itself. It wasn't until the mid-twentieth century that we realised the Norfolk Broads weren't natural lakes at all, but had been created by medieval peat extraction, with the depressions eventually claimed by rising sea levels. Rather than just proving something is anthropogenic, which we know *all* of Britain is in some way, we can use information gleaned from the past to understand changes in the character of the landscape over time, using data from soil cores, peat and wet places. This has the capacity to tell us how the landscape responded to particular interventions, what pollens became predominant, for instance, or how things recovered after a burning or farming regime, and can help to give some idea of the processes that might commence in different scenarios. There are also clues hidden in field names and old maps that can tell us about the historical tendency for somewhere to flood, or what the land may have been used for. It is not a panacea – the conditions now are not the same as they were – but it is informative nonetheless.

Archaeology is not a basis for the recreation of something that existed long ago. It is not possible to replicate all of the relationships and entanglements that went into its existence, and at any rate, those conditions were just as alive and changeable as they are now. To attempt to simulate a point in the past is surely more about making us *feel* like we're undertaking some kind of reparative justice in the face of our ecological grief, not necessarily about what might be best for now and the future. We are armed with more information than ever before about the world, our place in it and our impact on it. We need to use that to make our surroundings ecologically richer and more sustainable, not paper over the cracks. We cannot give in to demands that have been dreamt up by nostalgia, guilt and ideas of what is beautiful. Archaeology shows us that these islands of ours have taken thousands of years to shape and that what we do now will reverberate in ways that are probably unthinkable at present. It is for this reason we must look ahead to inform the legacy we want to leave, not try and invent the legacy we feel entitled to have inherited. We can't write places off and allow their potential to be traded for the promise of something down the line. If we can let go of what we wish the past had given to our present, maybe we can build a future where we embrace our place in nature rather than shun it. By weaving ourselves back into the wild with ways of living that serve both humans and the wild, with places where both can thrive and neither is the enemy of the other, perhaps what emerges will be a little more *Satochi-satoyama* and a little less *Jurassic Park*.

CONCLUSION
THE END

The real wild

My desk overlooks my rather messy garden; I didn't fully realise that I'm not a gardener until I owned it. When we first moved here the whole plot was overgrown with nettles and we did a decent job that first year of reclaiming it as something we could at least access. After digging and seeding grass over the bulk of it, the following summer we were surprised to find a mass of poppies growing on one side, some with huge red heads and others with fat pink pompoms. The poppies grew because we woke up the seeds as we disturbed the soil. How long they'd lain dormant I've no idea – the previous owner had passed away some years before we came to buy it. Try as we might, we were never able to replicate the beauty of that first full year in the garden. The poppies were a flash of someone else's gardening prowess. Perhaps one day when we're gone new owners will get this glimpse of the garden's former life, but I suspect for us it's over now.

As I wrote this book I paid more attention to observing what goes on outside my office window: what's coming and going, growing or diminishing. Dominant is an old pear tree, the last representative of a communal orchard that existed here from at least 1840, but probably earlier. Along the boundary is an edible hedgerow of dog rose, hawthorn, hazel and damson. We have a few small box shrubs, a lot of ground elder, and in a corner behind the shed a neighbour's bamboo is rapidly making itself at home. At one end is what we call 'the park', an area under enormous beech trees where the children have a tree swing and a sunken trampoline. Because it's a garden it's obviously a human-made thing, but now the staggering number of decisions and events that led to its emergence is more obvious. The pear, damson and ground elder are all surviving Roman introductions; the South East Asian bamboo is a Victorian legacy, along with a large *Fatus japonicus* and, I am sure, many other plants that show up occasionally. One fence line is all woven hazel, made the same way since prehistory. The beeches in the park are so dense they can only be a grown-out hedgerow which has formed that boundary for centuries. Alongside those is a footpath squashed between us and the older farmhouse next door that leads in one direction to the medieval common fields, in the other to a network of hidden paths meandering towards springs nestled in the folds of the chalk. The church here is highly likely to conceal earlier evidence of the veneration of one of the springs that bubbles out of a small cave in the greensand, set down millions of years ago.

It's a place that has been settled for millennia. It is not wild, but neither is the big stretch of grassland beyond that forms Salisbury Plain. There is wild*ness*, not the same as the Mesolithic or the medieval, but made up of what we have today. Grey squirrels, descendants of Victorian introductions from North America, chase each other up and down the pear tree and the beech, using the canopies as their own highway of safety from the neighbours' cats. Sometimes muntjacs, the miniature deer that are colonising Britain from their origin 160km away at Woburn, sneak in, and there are nocturnal visits from the fox and badger. In autumn we saw a goshawk bring down a pigeon and spend twenty minutes plucking the carcass in readiness for flying away with it. I've seen kingfishers in the band of scraggy woods by an old mill leat, and there is watercress going feral in the beds where the 'Vikings' were found. Still, the idea of places for us and places for them persists. A few months ago, a photo of a deer in someone's garden was posted on a local social media group and the first comment was, 'Poor things, running out of their own space.'

We have been divorcing ourselves from the wild since those earliest days when we started to manipulate clearings in the woods, repeating stories of danger and borrowing animal identities to ensure our success. As this led to a taming of the beasts, we also tamed some of our surroundings, learning to work things as we shifted the focus of our lives from survival to surplus. Settling into this new way of life meant redesigning on a landscape scale, creating new boundaries and edges that made Britain itself into a mosaic. With Roman rule came the notion of land ownership, and culinary and sporting tastes that required introductions into gardens and menageries, some of which later diffused to become part of the backdrop to everyday life. Wild things served the upper classes and were exploited for entertainment and status. While there was perhaps some reversion to a veneration for or a specialist approach to the wild in the early medieval period, 'wilderness' also became a place that was sought after for purity and solitude in burgeoning Christian narratives. With the Norman Conquest the earlier Roman approach to conferring status through the acquisition and control of wildlife was reinforced, though in Britain there was now very little, if any, 'true wild' left. No large predators remained and folkloric stories of the wild came to us as tales of resistance or as parables for the barbarism of the 'other'. By the eighteenth century, advances in science led us to dissect the world into neatly ordered categories of life, all while our landscape was so unrecognisable that people sought to recreate nature, now based on fiction, in which to escape. As Queen Victoria took to the throne our rose-tinted glasses were firmly in place and we saw wilderness where there was none, moving to protect what looked nice for the good of an increasingly urban nation. By the twenty-first century we were anxious about the effect we'd had on the world but no longer understood what a wild world looked like and failed to recognise the wildness in the places we'd been, because it didn't fit with what we thought it ought to be. We sought to alleviate our grief by finding anywhere that wasn't human-created, to prove to ourselves that nature could heal. Now, we seek to 'rewild' as if the answer to our problems lies in ceasing to exist, or turning back the clock to some point when everything was better.

Despite all of this, Britain is not an ecological wasteland. The wild has found a new way, and maybe we should be helping this on its way instead of clearing the board and starting again. Our canals have become refuges for trout and pike, and the road verges can be havens for pollinators. Grassland that may not have existed had we not cleared the forests harbours hares and lapwing, and the management of woodland rides can encourage woodcock that walk the paths picking up worms. Magical will o' the wisps may never even have been wild in the traditional sense in the first place, entirely made by the interplay between us and environmental conditions. Maybe without our fire they'll fade into pure folklore and will be imagined in the future as an allegory or in the same way we think of the Star Carr headdresses and the Lepaa artefact – stories of an older world and an older way of life that we can't quite grasp.

In the now, wildness is everywhere. We didn't leave it in the past and it was never really something separated from us. I don't believe it's useful to strive for something that no longer exists, and in some ways never did. We are integral to the ecology and biodiversity of our land. Its very structure and appearance has been shaped by what we've done, what we're doing and what we've left behind. The sheer diversity around us every single day is testament to millions of human lives lived, and the cumulative effect of countless decisions and accidents. What has prevailed isn't always what we intended and sometimes a seemingly innocuous choice has reverberated into the present and will continue into the future. In some there are opportunities: hedgerow forage or new sources of meat in introduced species. Others, such as invasive species, have been more detrimental. They don't even necessarily come from afar and aren't universally as obviously damaging as plants like knotweed, either.

We need to allow space for the fact that our past isn't always something to be proud of and that it has had effects we don't like, but we must recognise that we can't fix it by telling ourselves an overgrazed hillside is the pinnacle of wild beauty or that all trace of us needs to be erased. Our landscape can be so much more than it is, but requires us to negotiate the tricky terrain between our aspirations and reality. Looking to the past doesn't necessarily provide all the answers but it does give us vital context, and also shows us the things we have done *right.* It is evident in the animals and plants that have woven themselves into anthropogenic habitats – the bats in the mines and the bees in the buildings. Edgelands at the roadside, of rosehip, cow parsley and meadowsweet, provide shelter for the fieldmouse, pollen for the bees and edible crops for us, if we choose to look. Ignoring all of this is to ignore thousands of years of our relationship with nature and blind us to what we already have in pursuit of what we think we want.

Reimagined wilds

It is time to revisit the last wild places in Britain, the ones I asked you to imagine at the beginning of the book. I wonder if anything has changed, whether you go somewhere else or see people and their marks on the land and in the plants and animals around? Maybe the woods have old coppice stools filled with fungi and deadwood-loving beetles, or the ponies in the gorse on the heath have piqued an interest in their very cultural history. You may look for particular plants that enjoy disturbed ground, or wonder if there are bats in the old mines or nightjars in the scrubs. I hope that wherever it is, you will see that there is value in the wildness that carries our mark, even if it isn't always beautiful in that very traditional and pastoral sense. Rather than working on a series of trade-offs, trying to wipe ourselves from the map, let's break free of what we've been told is worthy and find the courage to look after the wild of the gaps. Solutions cannot be found in the mass abandonment of land any more than they can be found in mass development. In this mosaic we have made, however much it may sometimes resemble the bizarre wolf of Aldborough, each square contributes to the whole even though each has its own unique set of entwinements and is a manifestation of our role in the wider ecosystem. The next chapter in this story is ours to make, and it is only by accepting everything we are and have done within the web of wild things that we can ensure its survival – and ours – into the future.

NOTES

Full bibliographical details can be found in the Bibliography (p. 263).

1 Being with the Wild

1 Little et al., 'Technological Analysis of the World's Earliest Shamanic Costume'.
2 Conneller, *The Mesolithic in Britain.*
3 Warren et al., 'The Potential Role of Humans in Structuring the Wooded Landscapes of Mesolithic Ireland'.
4 Milner, Taylor and Conneller, 'Palaeoenvironmental Investigations'.
5 Barredo et al., 'Mapping and Assessment of Primary and Old-Growth Forests in Europe'.
6 Stereńczak et al., 'ALS-Based Detection of Past Human Activities in the Białowieża Forest – New Evidence of Unknown Remains of Past Agricultural Systems'.
7 Bailey et al., *The Archaeology of Europe's Drowned Landscapes.*
8 Hudson et al., 'Life before Stonehenge'.
9 Rageot et al., 'Birch Bark Tar Production'.
10 Momber et al., 'Mesolithic Occupation at Bouldnor Cliff and the Submerged Prehistoric Landscapes of the Solent'.
11 Finlay et al., 'Calling Time on Oronsay'.
12 Murray, Murray and Fraser, *A Tale of the Unknown Unknowns.*
13 Mithen, 'Mesolithic Fireplaces and the Enculturation of Early Holocene Landscapes in Britain, with a Case Study from Western Scotland'.
14 Cobb and Gray Jones, 'Being Mesolithic in Life and Death'.
15 Van Der Plicht et al., 'Surf'n Turf in Doggerland'.
16 Simonova, 'The Wild at Home and the Magic of Contact. Stories about Wild Animals and Spirits from Amudisy Evenki Hunters and Reindeer Herders'.
17 Orschiedt et al., 'The Shaman and the Infant'.
18 Cullon, 'A View from the Watchman's Pole'.
19 Nadasdy, 'The Gift in the Animal'.
20 Benavides Medina, 'The Insolent Fox'.
21 Wiessner, 'Embers of Society'.
22 Ahola and Lassila, 'Mesolithic Shadow Play?'

2 Taming

1 Gron et al., 'A Meeting in the Forest'.
2 Serjeantson, 'Survey of Animal Remains from Southern Britain Finds No Evidence for Continuity from the Mesolithic Period'.
3 Jones et al., 'The Origins of Agriculture'.
4 Chessa et al., 'Revealing the History of Sheep Domestication Using Retrovirus Integrations'.
5 Charlton et al., 'New Insights into Neolithic Milk Consumption through Proteomic Analysis of Dental Calculus'.
6 Serjeantson and Morris, 'Ravens and Crows in Iron Age and Roman Britain'.
7 Cochrane and Jones, *Visualising the Neolithic.*
8 Holmes, 'King of the Birds!'
9 Stanton, Mulville and Bruford, 'Colonization of the Scottish Islands via Long-Distance Neolithic Transport of Red Deer (*Cervus elaphus*)'.
10 *Tribeca Ltd* v. *Joules Limited* (2013).
11 Murphy and Ameen, 'The Shifting Baselines of the British Hare Goddess'.
12 Fyfe and Woodbridge, 'Differences in Time and Space in Vegetation Patterning'.
13 Vervust et al., 'Optically Stimulated Luminescence Profiling and Dating of Earthworks'.
14 Brown et al., 'Early to Middle Bronze Age Agricultural Terraces in North-East England'.
15 Bell and Leary, 'Pathways to Past Ways'.
16 Brennand et al., 'The Survey and Excavation of a Bronze Age Timber Circle at Holme-next-the-Sea, Norfolk, 1998–9'.
17 Chapman and Gearey, 'Iconoclasm in European Prehistory?'
18 Evans, Pollard and Knight, 'Life in the Woods'.
19 O'Sullivan, 'Exploring Past People's Interactions with Wetland Environments in Ireland'.
20 Hutton, 'Why Does Lindow Man Matter?'
21 Edwards, '*Will-o'-the-Wisp*'.
22 Mills, 'Will-o'-the-Wisp Revisited'.

3 Exoticism

1 Mattingly, *An Imperial Possession.*
2 Tomlin, *Roman London's First Voices.*
3 Cool, *Eating and Drinking in Roman Britain.*
4 Smout, *Nature Contested.*
5 Manning, Birley and Tipping, 'Roman Impact on the Environment at Hadrian's Wall'.

6 Cahill, 'Romans and Roman Material in Ireland'.
7 De Vareilles et al., 'The Development of Arable Cultivation in the South-East of England and Its Relationship with Vegetation Cover'.
8 Tomlin, 'A Five-Acre Wood in Roman Kent'.
9 Ferraby et al., *Thwing, Rudston and the Roman-Period Exploitation of the Yorkshire Wolds.*
10 Jarman et al., 'DNA Analysis of *Castanea sativa* (Sweet Chestnut) in Britain and Ireland'.
11 Neal, Wardle and Hunn, *Excavation of the Iron Age, Roman and Medieval Settlement at Gorhambury, St Albans.*
12 Kenward and Whitehouse, 'Insects'.
13 Walker, Sharpe and Williams, 'Barn Owls and Black Rats from a Rural Roman Villa at Gatehampton, South Oxfordshire'.
14 Trouwborst, McCormack and Martínez Camacho, 'Domestic Cats and Their Impacts on Biodiversity'.
15 Murphy and Ameen, 'The Shifting Baselines of the British Hare Goddess'.
16 Hughes, 'Mencius' Prescriptions for Ancient Chinese Environmental Problems'.
17 Parejko, 'Pliny the Elder's Silphium'.
18 Dambrine et al., 'Present Forest Biodiversity Patterns in France Related to Former Roman Agriculture'.

4 Control

1 Scott-Macnab, '"The Hunttyng of the Hare" in the Heege Manuscript'. The translation from Middle English is my own rendering.
2 Rackham, *The History of the Countryside.*
3 Dugdale, *The history of imbanking and drayning of divers fenns and marshes.*
4 Rippon, *The Transformation of Coastal Wetlands.*
5 Webb and Goodenough, 'Questioning the Reliability of "Ancient" Woodland Indicators'.
6 Mitchell, 'The Development of Ireland's Tree Cover over the Millennia'.
7 Tallis and Switsur, 'Studies on Southern Pennine Peats'.
8 Hooke, 'Pre-Conquest Woodland'.
9 Liddiard, 'The Deer Parks of Domesday Book'.
10 Cantor and Hatherley, 'The Medieval Parks of England'.
11 Sykes and Ameen, 'Preliminary Report on the Animal Bones from Clarendon Palace'.
12 O'Regan, 'Brown Bears in Burials and Entertainment in Later Prehistoric to Modern Britain (*c.*2400 BC–AD 1900s)'.
13 Barrett, Locker and Roberts, 'The Origins of Intensive Marine Fishing in Medieval Europe'.

14 O'Connor and Sykes, *Extinctions and Invasions.*
15 Whitaker, 'Sarsen Stone Quarrying in Southern England'.
16 Semple, 'A Fear of the Past'.
17 Moore, 'Myths and Folklore as Aids in Interpreting the Prehistoric Landscape at the Carrowkeel Passage Tomb Complex, Co. Sligo, Ireland'.
18 Hutton, 'The Wild Hunt and the Witches' Sabbath'.
19 Pollard, *Imagining Robin Hood.*
20 Pluskowski, *Wolves and the Wilderness in the Middle Ages.*

5 Dissection

1 Williamson, *An Environmental History of Wildlife in England 1650–1950.*
2 Schulte, 'The Werewolf in the Popular Culture of Early Modern Germany'.
3 Pemberton, Strange and Worboys, 'Breeding and Breed'.
4 Miller, *An Environmental History of Latin America.*
5 Koch et al., 'Earth System Impacts of the European Arrival and Great Dying in the Americas after 1492'.
6 Rhodes, 'Newhailes'.
7 Rattue, 'Gardens and Designed Landscapes'.
8 Lowe and Gardiner, 'A Re-examination of the Subspecies of Red Deer (*Cervus elaphus*) with Particular Reference to the Stocks in Britain'.
9 Putman, 'The New Forest and Its Larger Herbivores'.
10 Hay, *Albion's Fatal Tree.*
11 Ezban, 'Decoys, Dikes and Lures'.

6 Protection

1 Wordsworth, *A guide through the district of the lakes in the north of England: with a description of the scenery, &c., for the use of tourists and residents.*
2 Braungardt et al., 'Arsenic Concentrations, Distributions and Bioaccessibilities at a UNESCO World Heritage Site (Devon Great Consols, Cornwall and West Devon Mining Landscape)'.
3 Cox and Rodway-Dyer, 'The Underappreciated Value of Brownfield Sites'.
4 Hickman, '"To Brighten the Aspect of Our Streets and Increase the Health and Enjoyment of Our City"'.
5 Taylor, *A Claim on the Countryside.*
6 Robinson and Sutherland, 'Post-war Changes in Arable Farming and Biodiversity in Great Britain'.
7 Sainsbury et al., 'Recent History, Current Status, Conservation and Management of Native Mammalian Carnivore Species in Great Britain'.
8 Cassidy, *Vermin, Victims and Disease.*
9 Lydekker, *The Deer of All Lands.*

10 Whitehead, 'The Muntjac in Britain'.
11 Ward, Richardson and Mergeay, 'Reeves' Muntjac Populations Continue to Grow and Spread across Great Britain and Are Invading Continental Europe'.
12 Du Plessis et al., 'Genomics Reveals Complex Population History and Unexpected Diversity of Eurasian Otters (*Lutra lutra*) in Britain Relative to Genetic Methods'.
13 Costa et al., 'The Genetic Legacy of the 19th-Century Decline of the British Polecat'.
14 Urban et al., 'Ancient *Mycobacterium leprae* Genome Reveals Medieval English Red Squirrels as Animal Leprosy Host'.
15 Anon, 'Modern Legends of Supposed "Wolves" in Epping Forest'.
16 Steinhart, *The Company of Wolves*.
17 Murugesu, 'Is There Really a Big Cat in the Lake District?'
18 Coard, 'Ascertaining an Agent'.
19 Mullan and Marvin, *Zoo Culture*.
20 Thompsell, *Hunting Africa*.
21 Anon., 'Przevalsky's Wild Horse'.
22 Orlando, 'Ancient Genomes Reveal Unexpected Horse Domestication and Management Dynamics'.
23 Der Sarkissian et al., 'Evolutionary Genomics and Conservation of the Endangered Przewalski's Horse'.
24 Hiemstra et al., 'Bird Nests Made from Anti-Bird Spikes'.

7 The New Wild

1 Glentworth, Gilchrist and Avery, 'The Place for People in Rewilding'.
2 Wiener and Wilkinson, 'Deciphering the Genetic Basis of Animal Domestication'.
3 Lysaght, 'Heaney vs Praeger'.
4 'Statutory Biodiversity Credit Prices'.
5 Zu Ermgassen et al., 'Exploring the Ecological Outcomes of Mandatory Biodiversity Net Gain Using Evidence from Early-Adopter Jurisdictions in England'.
6 Orros and Fellowes, 'Widespread Supplementary Feeding in Domestic Gardens Explains the Return of Reintroduced Red Kites'.
7 Fitzgerald, 'A Social History of the Slaughterhouse'.
8 'Creutzfeldt-Jakob Disease in the UK (by Calendar Year)'.
9 Hassanin, Grandcolas and Veron, 'Covid-19'; Ray and Bhattacharya, 'An Overview on the Zoonotic Aspects of COVID-19'.
10 Bengis et al., 'The Role of Wildlife in Emerging and Re-Emerging Zoonoses'.

11 Fournié, Pfeiffer and Bendrey, 'Early Animal Farming and Zoonotic Disease Dynamics'.
12 Rayfield et al., 'Uncovering the Holocene Roots of Contemporary Disease-Scapes'.
13 Morand, McIntyre and Baylis, 'Domesticated Animals and Human Infectious Diseases of Zoonotic Origins'.

BIBLIOGRAPHY

Ahola, Marja and Katri Lassila, 'Mesolithic Shadow Play? Exploring the Performative Attributes of a Zoomorphic Wild Reindeer (*Rangifer tarandus*) Antler Artefact from Finland'. *Time and Mind* 15, no. 2 (3 April 2022): 167–85. https://doi.org/10.1080/1751696X.2022.2098047.

Anon., 'Modern Legends of Supposed "Wolves" in Epping Forest'. *Journals of Proceedings of the Essex Field Club* 4 (1884): cciv–ccix.

Anon., 'Przevalsky's Wild Horse'. *Nature* 30, no. 773 (August 1884): 391–2. https://doi.org/10.1038/030391a0.

Bailey, Geoff, Nena Galanidou, Hans Peeters, Hauke Jöns and Moritz Mennenga, eds, *The Archaeology of Europe's Drowned Landscapes*. Vol. 35. Coastal Research Library. Cham: Springer International Publishing, 2020. https://doi.org/10.1007/978-3-030-37367-2.

Barredo, José, Cristina Brailescu, Anne Teller, Francesco Maria Sabatini, Achille Mauri and Klara Janouskova, 'Mapping and Assessment of Primary and Old-Growth Forests in Europe'. JRC Science for Policy Report. Luxembourg City: European Commission, 2021.

Barrett, James H., Alison M. Locker and Callum M. Roberts, 'The Origins of Intensive Marine Fishing in Medieval Europe: The English Evidence'. *Proceedings of the Royal Society of London. Series B: Biological Sciences* 271, no. 1556 (7 December 2004): 2417–21. https://doi.org/10.1098/rspb.2004.2885.

Bell, Martin and Jim Leary, 'Pathways to Past Ways: A Positive Approach to Routeways and Mobility'. *Antiquity* 94, no. 377 (October 2020): 1349–59. https://doi.org/10.15184/aqy.2020.133.

Benavides Medina, Sebastián Pelayo, 'The Insolent Fox: Human–Animal Relations with Protected Predators in Central-Southern Chile'. *Anthrozoös* 33, no. 5 (2 September 2020): 597–612. https://doi.org/10.1080/08927936.2020.1799547.

Bengis, R. G., F. A. Leighton, J. R. Fischer, M. Artois, T. Mörner and C. M. Tate, 'The Role of Wildlife in Emerging and Re-Emerging Zoonoses'. *Rev. Sci. Tech.* 23, no. 2 (August 2004): 497–511.

Braungardt, Charlotte, Xiaqing Chen, Daniel Chester-Sterne, James G. A. Quinn and Andrew Turner, 'Arsenic Concentrations, Distributions and Bioaccessibilities at a UNESCO World Heritage Site (Devon Great Consols, Cornwall and West Devon Mining Landscape)'. *Environmental Pollution* 264 (September 2020): 114590. https://doi.org/10.1016/j.envpol.2020.114590.

Brennand, Mark, Maisie Taylor, Trevor Ashwin, Alex Bayliss, Matt Canti, Andrew Chamberlain, C. A. I. French et al., 'The Survey and Excavation of a Bronze Age Timber Circle at Holme-next-the-Sea, Norfolk, 1998–9'. *Proceedings of the Prehistoric Society* 69 (2003): 1–84. https://doi.org/10.1017/S0079497X00001250.

Brown, Antony G., Daniel Fallu, Sara Cucchiaro, Monica Alonso-Eguiluz, Rosa Maria Albert, Kevin Walsh, Ben R. Pears et al., 'Early to Middle Bronze Age Agricultural Terraces in North-East England: Morphology, Dating and Cultural Implications'. *Antiquity* 97, no. 392 (April 2023): 348–66. https://doi.org/10.15184/aqy.2023.1.

Cahill, Jacqueline, 'Romans and Roman Material in Ireland: A Wider Social Perspective'. In *Late Iron Age and 'Roman' Ireland*, edited by The Discovery Programme Limited, Discovery Programme Reports 8, 11–58. Dublin: Wordwell, 2014.

Cantor, L. M. and J. Hatherley, 'The Medieval Parks of England'. *Geography* 64, no. 2 (1979): 71–85.

Cassidy, Angela, *Vermin, Victims and Disease: British Debates over Bovine Tuberculosis and Badgers*. Cham: Springer International Publishing, 2019. https://doi.org/10.1007/978-3-030-19186-3.

Chapman, Henry and Benjamin Gearey, 'Iconoclasm in European Prehistory? Breaking Objects and Landscapes'. In *Striking Images, Iconoclasms Past and Present*, edited by Stacy Boldrick, Leslie Brubaker and Richard Clay, 25–38. Burlington, VT: Ashgate, 2013.

Charlton, Sophy, Abigail Ramsøe, Matthew Collins, Oliver E. Craig, Roman Fischer, Michelle Alexander and Camilla F. Speller, 'New Insights into Neolithic Milk Consumption through Proteomic Analysis of Dental Calculus'. *Archaeological and Anthropological Sciences* 11, no. 11 (November 2019): 6183–96. https://doi.org/10.1007/s12520-019-00911-7.

Chessa, Bernardo, Filipe Pereira, Frederick Arnaud, Antonio Amorim, Félix Goyache, Ingrid Mainland, Rowland R. Kao et al., 'Revealing the History of Sheep Domestication Using Retrovirus Integrations'. *Science* 324, no. 5926 (24 April 2009): 532–6. https://doi.org/10.1126/science.1170587.

Coard, Ros, 'Ascertaining an Agent: Using Tooth Pit Data to Determine the Carnivore/s Responsible for Predation in Cases of Suspected Big Cat Kills in an Upland Area of Britain'. *Journal of Archaeological Science* 34, no. 10 (October 2007): 1677–84. https://doi.org/10.1016/j.jas.2006.12.006.

Cobb, Hannah and Amy Gray Jones, 'Being Mesolithic in Life and Death'. *Journal of World Prehistory* 31, no. 3 (September 2018): 367–83. https://doi.org/10.1007/s10963-018-9123-1.

Cochrane, Andrew and Andrew Meirion Jones, eds, *Visualising the Neolithic*. Oxford: Oxbow Books, 2012. https://doi.org/10.2307/j.ctvh1dwd6.

Conneller, Chantal, *The Mesolithic in Britain: Landscape and Society in Times of Change*. Routledge Archaeology of Northern Europe. Abingdon: Routledge, 2022.

Cool, H. E. M., *Eating and Drinking in Roman Britain*. Cambridge: Cambridge University Press, 2006.

Costa, M., C. Fernandes, J. D. S. Birks, A. C. Kitchener, M. Santos-Reis and M. W. Bruford, 'The Genetic Legacy of the 19th-Century Decline of the British Polecat: Evidence for Extensive Introgression from Feral Ferrets'. *Molecular Ecology* 22, no. 20 (October 2013): 5130–47. https://doi.org/10.1111/mec.12456.

Cox, Lauren and Sue Rodway-Dyer, 'The Underappreciated Value of Brownfield Sites: Motivations and Challenges Associated with Maintaining Biodiversity'. *Journal of Environmental Planning and Management* 66, no. 9 (29 July 2023): 2009–27. https://doi.org/10.1080/09640568.2022.2050683.

'Creutzfeldt-Jakob Disease in the UK (by Calendar Year)'. The National CJD Research & Surveillance Unit (NCJDRSU), 2023. https://www.cjd.ed.ac.uk/sites/default/files/figs.pdf.

Cullon, Deidre, 'A View from the Watchman's Pole: Salmon, Animism and the Kwakwaka'wakw Summer Ceremonial'. *BC Studies: The British Columbian Quarterly* (15 April 2013): 9–37. https://doi.org/10.14288/BCS.V0I177.182922.

Dambrine, E., J.-L. Dupouey, L. Laüt, L. Humbert, M. Thinon, T. Beaufils and H. Richard, 'Present Forest Biodiversity Patterns in France Related to Former Roman Agriculture'. *Ecology* 88, no. 6 (June 2007): 1430–9. https://doi.org/10.1890/05-1314.

De Vareilles, Anne, Jessie Woodbridge, Ruth Pelling, Ralph Fyfe, David Smith, Gill Campbell, Wendy Smith et al., 'The Development of Arable Cultivation in the South-East of England and Its Relationship with Vegetation Cover: A Honeymoon Period for Biodiversity?' *The Holocene* 33, no. 11 (November 2023): 1389–406. https://doi.org/10.1177/09596836231185836.

Der Sarkissian, Clio, Luca Ermini, Mikkel Schubert, Melinda A. Yang, Pablo Librado, Matteo Fumagalli, Hákon Jónsson et al., 'Evolutionary Genomics and Conservation of the Endangered Przewalski's Horse'. *Current Biology* 25, no. 19 (October 2015): 2577–83. https://doi.org/10.1016/j.cub.2015.08.032.

Du Plessis, Sarah J., Mark Blaxter, Klaus-Peter Koepfli, Elizabeth A. Chadwick and Frank Hailer, 'Genomics Reveals Complex Population History and Unexpected Diversity of Eurasian Otters (*Lutra lutra*) in Britain Relative to Genetic Methods'. Edited by Emma Teeling. *Molecular Biology and Evolution* 40, no. 11 (3 November 2023): msad207. https://doi.org/10.1093/molbev/msad207.

Dugdale, William, *The history of imbanking and drayning of divers fenns and marshes: both in forein parts and in this kingdom: and of the improvements thereby: extracted from records, manuscripts, and other authentick testimonies.* 1662. http://galenet.galegroup.com/servlet/MOME?af=RN&ae=U100236060&srchtp=a&ste=14&locID=sfu_z39.

Edwards, Howell G. M., '*Will-o'-the-Wisp*: An Ancient Mystery with Extremophile Origins?' *Philosophical Transactions of the Royal Society A: Mathematical, Physical and Engineering Sciences* 372, no. 2030 (13 December 2014): 20140206. https://doi.org/10.1098/rsta.2014.0206.

Evans, Christopher, Joshua Pollard and Mark Knight, 'Life in the Woods: Tree-throws, "Settlement" and Forest Cognition'. *Oxford Journal of Archaeology* 18, no. 3 (August 1999): 241–54. https://doi.org/10.1111/1468-0092.00081.

Ezban, Michael, 'Decoys, Dikes and Lures: Polyfunctional Landscapes of Waterfowl Hunting'. *Studies in the History of Gardens & Designed Landscapes* 33, no. 3 (2013): 193–207. https://doi.org/10.1080/14601176.2013.820921.

Ferraby, Rose, Paul Johnson, Martin Millet and Lacey M. Wallace, eds, *Thwing, Rudston and the Roman-Period Exploitation of the Yorkshire Wolds*. Yorkshire Archaeological Report 8. Leeds: Yorkshire Archaeological and Historical Society, 2017.

Finlay, Nyree, Ruby Cerón-Carrasco, Rupert Housley, Jeremy Huggett, W. Graham Jardine, Susan Ramsay, Catherine Smith, Dene Wright, Julian Augley and Peter J. Wright, 'Calling Time on Oronsay: Revising Settlement Models Around the Mesolithic–Neolithic Transition in Western Scotland, New Evidence from Port Lobh, Colonsay'. *Proceedings of the Prehistoric Society* 85 (December 2019): 83–114. https://doi.org/10.1017/ppr.2019.2.

Fitzgerald, Amy J., 'A Social History of the Slaughterhouse: From Inception to Contemporary Implications'. *Human Ecology Review* 17, no. 1 (2010): 58–69.

Fournié, Guillaume, Dirk U. Pfeiffer and Robin Bendrey, 'Early Animal Farming and Zoonotic Disease Dynamics: Modelling Brucellosis Transmission in Neolithic Goat Populations'. *Royal Society Open Science* 4, no. 2 (February 2017): 160943. https://doi.org/10.1098/rsos.160943.

Fyfe, Ralph M. and Jessie Woodbridge, 'Differences in Time and Space in Vegetation Patterning: Analysis of Pollen Data from Dartmoor, UK'. *Landscape Ecology* 27, no. 5 (May 2012): 745–60. https://doi.org/10.1007/s10980-012-9726-3.

Glentworth, Joseph, Anna Gilchrist and Rowan Avery, 'The Place for People in Rewilding'. *Conservation Biology*, July 2024, e14318. https://doi.org/10.1111/cobi.14318.

Gron, Kurt J., Peter Rowley-Conwy, Eva Fernandez-Dominguez, Darren R. Gröcke, Janet Montgomery, Geoff M. Nowell and William P. Patterson, 'A Meeting in the Forest: Hunters and Farmers at the Coneybury "Anomaly", Wiltshire'. *Proceedings of the Prehistoric Society* 84 (December 2018): 111–44. https://doi.org/10.1017/ppr.2018.15.

Hassanin, Alexandre, Philippe Grandcolas and Géraldine Veron, 'Covid-19: Natural or Anthropic Origin?' *Mammalia* 85, no. 1 (27 January 2021): 1–7. https://doi.org/10.1515/mammalia-2020-0044.

Hay, Douglas, ed., *Albion's Fatal Tree: Crime and Society in Eighteenth-Century England*. New York: Pantheon Books, 1975.

Hickman, Clare, '"To Brighten the Aspect of Our Streets and Increase the Health and Enjoyment of Our City": The National Health Society and Urban Green

Space in Late-Nineteenth Century London'. *Landscape and Urban Planning* 118 (October 2013): 112–19. https://doi.org/10.1016/j.landurbplan.2012.09.007.

Hiemstra, Auke-Florian, Cornelis W. Moeliker, Barbara Gravendeel and M. Schilthuizen, 'Bird Nests Made from Anti-Bird Spikes'. *Deinsea* 21 (2023): 17–25.

Holmes, Matilda, 'King of the Birds! The Changing Role of White-Tailed (*Haliaeetus albicilla*) and Golden-Eagles (*Aquila chrysaetos*) in Britain's Past'. *Archaeofauna* 27 (23 November 2018): 173–94. https://doi.org/10.15366/archaeofauna2018.27.007.

Hooke, Della, 'Pre-Conquest Woodland: Its Distribution and Usage'. *The Agricultural History Review* 37, no. 2 (1989): 113–29.

Hudson, Samuel M., Ben Pears, David Jacques, Thierry Fonville, Paul Hughes, Inger Alsos, Lisa Snape, Andreas Lang and Antony Brown, 'Life before Stonehenge: The Hunter-Gatherer Occupation and Environment of Blick Mead Revealed by sedaDNA, Pollen and Spores'. Edited by Peter F. Biehl. *PLOS ONE* 17, no. 4 (27 April 2022): e0266789. https://doi.org/10.1371/journal.pone.0266789.

Hughes, J. Donald, 'Mencius' Prescriptions for Ancient Chinese Environmental Problems'. *Environmental Review* 13, no. 3–4 (1 September 1989): 15–27. https://doi.org/10.2307/3984388.

Hutton, Ronald, 'Why Does Lindow Man Matter?' *Time and Mind* 4, no. 2 (January 2011): 135–48. https://doi.org/10.2752/175169711X12961583765171.

——— 'The Wild Hunt and the Witches' Sabbath'. *Folklore* 125, no. 2 (4 May 2014): 161–78. https://doi.org/10.1080/0015587X.2014.896968.

Jarman, Rob, Claudia Mattioni, Karen Russell, Frank M. Chambers, Debbie Bartlett, M. Angela Martin, Marcello Cherubini, Fiorella Villani and Julia Webb, 'DNA Analysis of *Castanea sativa* (Sweet Chestnut) in Britain and Ireland: Elucidating European Origins and Genepool Diversity'. Edited by Filippos A. Aravanopoulos. *PLOS ONE* 14, no. 9 (25 September 2019): e0222936. https://doi.org/10.1371/journal.pone.0222936.

Jones, Glynis, Thomas Kluyver, Catherine Preece, Jennifer Swarbrick, Emily Forster, Michael Wallace, Michael Charles, Mark Rees and Colin P. Osborne, 'The Origins of Agriculture: Intentions and Consequences'. *Journal of Archaeological Science* 125 (January 2021): 105290. https://doi.org/10.1016/j.jas.2020.105290.

Kenward, Harry and Nicki Whitehouse, 'Insects'. In *Extinctions and Invasions: A Social History of British Fauna*, edited by Terry O'Connor and Naomi Jane Sykes, 181–9. Oxford: Windgather Press, 2010.

Koch, Alexander, Chris Brierley, Mark M. Maslin and Simon L. Lewis, 'Earth System Impacts of the European Arrival and Great Dying in the Americas after 1492'. *Quaternary Science Reviews* 207 (March 2019): 13–36. https://doi.org/10.1016/j.quascirev.2018.12.004.

Liddiard, Robert, 'The Deer Parks of Domesday Book'. *Landscapes* 1 (2003): 4–23.

Little, Aimée, Benjamin Elliott, Chantal Conneller, Diederik Pomstra, Adrian A. Evans, Laura C. Fitton, Andrew Holland et al., 'Technological Analysis of the World's Earliest Shamanic Costume: A Multi-Scalar, Experimental Study of a Red Deer Headdress from the Early Holocene Site of Star Carr, North Yorkshire, UK'. Edited by Michael D. Petraglia. *PLOS ONE* 11, no. 4 (13 April 2016): e0152136. https://doi.org/10.1371/journal.pone.0152136.

Lowe, V. P. W. and A. S. Gardiner, 'A Re-examination of the Subspecies of Red Deer (*Cervus elaphus*) with Particular Reference to the Stocks in Britain'. *Journal of Zoology* 174, no. 2 (October 1974): 185–201. https://doi.org/10.1111/j.1469-7998.1974.tb03151.x.

Lydekker, Richard, *The Deer of All Lands; a History of the Family Cervidæ Living and Extinct*. London: R. Ward, limited, 1898. https://doi.org/10.5962/bhl.title.77310.

Lysaght, Seán, 'Heaney vs Praeger: Contrasting Natures'. *The Irish Review (1986–)*, no. 7 (1989): 68. https://doi.org/10.2307/29735473.

Manning, Adrian, Robin Birley and Richard Tipping, 'Roman Impact on the Environment at Hadrian's Wall: Precisely Dated Pollen Analysis from Vindolanda, Northern England'. *The Holocene* 7, no. 2 (June 1997): 175–86. https://doi.org/10.1177/095968369700700205.

Mattingly, David J., *An Imperial Possession: Britain in the Roman Empire, 54 BC–AD 409*. The Penguin History of Britain. London: Penguin Books, 2007.

Miller, Shawn William, *An Environmental History of Latin America*. New Approaches to the Americas. Cambridge: Cambridge University Press, 2007. https://doi.org/10.1017/CBO9780511800672.

Mills, A. A., 'Will-o'-the-Wisp Revisited'. *Weather* 55, no. 7 (July 2000): 239–41. https://doi.org/10.1002/j.1477-8696.2000.tb04067.x.

Milner, Nicky, Barry Taylor and Chantal Conneller, eds, 'Palaeoenvironmental Investigations'. In *Star Carr Volume II*, 123–49. York: White Rose University Press, 2018. https://doi.org/10.22599/book2.e.

Mitchell, Fraser, 'The Development of Ireland's Tree Cover over the Millennia'. *Irish Forestry Journal* 58 (2000): 38–46.

Mithen, Steven, 'Mesolithic Fireplaces and the Enculturation of Early Holocene Landscapes in Britain, with a Case Study from Western Scotland'. *Proceedings of the Prehistoric Society* 85 (December 2019): 131–59. https://doi.org/10.1017/ppr.2019.6.

Momber, Garry, David Tomalin, Rob Scaife, Julie Satchell and Jan Gillespie, 'Mesolithic Occupation at Bouldnor Cliff and the Submerged Prehistoric Landscapes of the Solent'. CBA Research Reports. York: Council for British Archaeology, 2011.

Moore, Sam, 'Myths and Folklore as Aids in Interpreting the Prehistoric Landscape at the Carrowkeel Passage Tomb Complex, Co. Sligo, Ireland'. In *Folk Beliefs and Practice in Medieval Lives*, edited by Ann-Britt Falk and Donata M. Kyritz. BAR 1757, 7–21. Oxford: Archaeopress, 2008.

Morand, Serge, K. Marie McIntyre and Matthew Baylis, 'Domesticated Animals and Human Infectious Diseases of Zoonotic Origins: Domestication Time Matters'. *Infection, Genetics and Evolution* 24 (June 2014): 76–81. https://doi.org/10.1016/j.meegid.2014.02.013.

Mullan, Bob and Garry Marvin, *Zoo Culture*. 2nd edn. Chicago, IL: University of Illinois Press, 1999.

Murphy, Luke John and Carly Ameen, 'The Shifting Baselines of the British Hare Goddess'. *Open Archaeology* 6, no. 1 (10 October 2020): 214–35. https://doi.org/10.1515/opar-2020-0109.

Murray, Hilary K., J. C. Murray and Caroline Fraser, *A Tale of the Unknown Unknowns: A Mesolithic Pit Alignment and a Neolithic Timber Hall at Warren Field, Crathes, Aberdeenshire*. Oxford: Oxbow Books, 2009.

Murugesu, Jason Arunn, 'Is There Really a Big Cat in the Lake District?' BBC News, 27 May 2024. https://www.bbc.co.uk/news/articles/cqvv25j8gx1o.

Nadasdy, Paul, 'The Gift in the Animal: The Ontology of Hunting and Human–Animal Sociality'. *American Ethnologist* 34, no. 1 (February 2007): 25–43. https://doi.org/10.1525/ae.2007.34.1.25.

Neal, David S., Angela Wardle and Jonathan R. Hunn, *Excavation of the Iron Age, Roman and Medieval Settlement at Gorhambury, St Albans*. Historic Buildings and Monuments Commission for England Archaeological Reports 14. London: English Heritage Historic Buildings and Monuments Commission for England, 1990.

O'Connor, Terence Patrick and Naomi Jane Sykes, *Extinctions and Invasions: A Social History of British Fauna*. Oxford: Windgather Press, 2010.

O'Regan, Hannah J., 'Brown Bears in Burials and Entertainment in Later Prehistoric to Modern Britain (*c.*2400 BC–AD 1900s)'. In *Bear and Human*, edited by Oliver Grimm, 187–208. Turnhout: Brepols Publishers, 2023. https://doi.org/10.1484/M.TANE-EB.5.134334.

Orlando, Ludovic, 'Ancient Genomes Reveal Unexpected Horse Domestication and Management Dynamics'. *BioEssays* 42, no. 1 (January 2020): 1900164. https://doi.org/10.1002/bies.201900164.

Orros, Melanie E. and Mark D. E. Fellowes, 'Widespread Supplementary Feeding in Domestic Gardens Explains the Return of Reintroduced Red Kites *Milvus milvus* to an Urban Area'. Edited by Jose Antonio Sanchez-Zapata. *Ibis* 157, no. 2 (April 2015): 230–8. https://doi.org/10.1111/ibi.12237.

Orschiedt, Jörg, Wolfgang Haak, Holger Dietl, Andreas Siegl and Harald Meller, 'The Shaman and the Infant: The Mesolithic Double Burial from Bad Dürrenberg, Germany'. In *Kinship, Sex, and Biological Relatedness*, edited by Harald Meller, Johannes Krause, Wolfgang Haak and Roberto Risch, 125–36. Heidelberg: Propylaeum, 2023. https://doi.org/10.11588/PROPYLAEUM.1280.C18002.

O'Sullivan, Aidan, 'Exploring Past People's Interactions with Wetland Environments in Ireland'. *Proceedings of the Royal Irish Academy: Archaeology,*

Culture, History, Literature 107C, no. 1 (2007): 147–203. https://doi.org/10.1353/ria.2007.0005.

Parejko, Ken, 'Pliny the Elder's Silphium: First Recorded Species Extinction'. *Conservation Biology* 17, no. 3 (June 2003): 925–7. https://doi.org/10.1046/j.1523-1739.2003.02067.x.

Pemberton, Neil, Julie-Marie Strange and Michael Worboys, 'Breeding and Breed'. In *The Routledge Companion to Animal–Human History*, edited by Hilda Kean and Philip Howell, 393–421. Abingdon, New York: Routledge, 2018.

Pluskowski, Aleksander, *Wolves and the Wilderness in the Middle Ages*. Woodbridge: Boydell Press, 2006.

Pollard, A. J., *Imagining Robin Hood*. Abingdon: Routledge, 2004. https://doi.org/10.4324/9780203005521.

Putman, Rory J., 'The New Forest and Its Larger Herbivores'. In *Competition and Resource Partitioning in Temperate Ungulate Assemblies*, by Rory J. Putman, 11–29. Dordrecht: Springer Netherlands, 1996. https://doi.org/10.1007/978-94-009-1517-6_2.

Rackham, Oliver, *The History of the Countryside: The Classic History of Britain's Landscape, Flora and Fauna*. London: Phoenix Press, 2000.

Rageot, Maxime, Isabelle Théry-Parisot, Sylvie Beyries, Cédric Lepère, Alain Carré, Arnaud Mazuy, Jean-Jacques Filippi, Xavier Fernandez, Didier Binder and Martine Regert, 'Birch Bark Tar Production: Experimental and Biomolecular Approaches to the Study of a Common and Widely Used Prehistoric Adhesive'. *Journal of Archaeological Method and Theory* 26, no. 1 (March 2019): 276–312. https://doi.org/10.1007/s10816-018-9372-4.

Rattue, James, 'Gardens and Designed Landscapes'. In *The Palgrave Handbook of Gothic Origins*, edited by Clive Bloom, 345–62. Cham: Springer International Publishing, 2021. https://doi.org/10.1007/978-3-030-84562-9_17.

Ray, Anushree Singha and Kuntal Bhattacharya, 'An Overview on the Zoonotic Aspects of COVID-19'. *Proceedings of the National Academy of Sciences, India Section B: Biological Sciences* 94, no. 1 (February 2024): 9–13. https://doi.org/10.1007/s40011-023-01445-8.

Rayfield, Kristen M., Alexis M. Mychajliw, Robin R. Singleton, Sabrina B. Sholts and Courtney A. Hofman, 'Uncovering the Holocene Roots of Contemporary Disease-Scapes: Bringing Archaeology into One Health'. *Proceedings of the Royal Society B: Biological Sciences* 290, no. 2012 (6 December 2023): 20230525. https://doi.org/10.1098/rspb.2023.0525.

Rhodes, Daniel T., 'Newhailes: An 18th-Century Designed Landscape in Scotland and Its Role in Enlightenment Social Theatre'. *Post-Medieval Archaeology* 55, no. 1 (2 January 2021): 15–38. https://doi.org/10.1080/00794236.2021.1894854.

Rippon, Stephen, *The Transformation of Coastal Wetlands: Exploitation and Management of Marshland Landscapes in North West Europe during the Roman and Medieval Periods*. A British Academy Postdoctoral Fellowship Monograph. Oxford: Oxford University Press, 2000.

Robinson, Robert A. and William J. Sutherland, 'Post-war Changes in Arable Farming and Biodiversity in Great Britain'. *Journal of Applied Ecology* 39, no. 1 (February 2002): 157–76. https://doi.org/10.1046/j.1365-2664.2002.00695.x.

Sainsbury, Katherine A., Richard F. Shore, Henry Schofield, Elizabeth Croose, Ruairidh D. Campbell and Robbie A. McDonald, 'Recent History, Current Status, Conservation and Management of Native Mammalian Carnivore Species in Great Britain'. *Mammal Review* 49, no. 2 (April 2019): 171–88. https://doi.org/10.1111/mam.12150.

Schulte, Rolf, 'The Werewolf in the Popular Culture of Early Modern Germany'. In *Werewolf Histories*, edited by Willem De Blécourt, 185–205. London: Palgrave Macmillan, 2015. https://doi.org/10.1007/978-1-137-52634-2_8.

Scott-Macnab, David, '"The Hunttyng of the Hare" in the Heege Manuscript'. *Anglia – Zeitschrift für Englische Philologie* 128, no. 1 (January 2010). https://doi.org/10.1515/angl.2010.009.

Semple, Sarah, 'A Fear of the Past: The Place of the Prehistoric Burial Mound in the Ideology of Middle and Later Anglo-Saxon England'. *World Archaeology* 30, no. 1 (June 1998): 109–26. https://doi.org/10.1080/00438243.1998.9980400.

Serjeantson, Dale, 'Survey of Animal Remains from Southern Britain Finds No Evidence for Continuity from the Mesolithic Period'. *Environmental Archaeology* 19, no. 3 (October 2014): 256–62. https://doi.org/10.1179/1749631414Y.0000000020.

Serjeantson, Dale and John Morris, 'Ravens and Crows in Iron Age and Roman Britain'. *Oxford Journal of Archaeology* 30, no. 1 (February 2011): 85–107. https://doi.org/10.1111/j.1468-0092.2010.00360.x.

Simonova, Veronika V., 'The Wild at Home and the Magic of Contact. Stories about Wild Animals and Spirits from Amudisy Evenki Hunters and Reindeer Herders'. *Études Mongoles et Sibériennes, Centrasiatiques et Tibétaines* 49 (2018). https://doi.org/10.4000/emscat.3505.

Smout, T. C., *Nature Contested: Environmental History in Scotland and Northern England since 1600*. Edinburgh: Edinburgh University Press, 2022. https://doi.org/10.1515/9781474472715.

Stanton, David W. G., Jacqueline A. Mulville and Michael W. Bruford, 'Colonization of the Scottish Islands via Long-Distance Neolithic Transport of Red Deer (*Cervus elaphus*)'. *Proceedings of the Royal Society B: Biological Sciences* 283, no. 1828 (13 April 2016): 20160095. https://doi.org/10.1098/rspb.2016.0095.

'Statutory Biodiversity Credit Prices'. Department for Environment, Food & Rural Affairs, 2023. https://www.gov.uk/guidance/statutory-biodiversity-credit-prices.

Steinhart, Peter, *The Company of Wolves*. New York: Knopf Doubleday Publishing Group, 2011.

Stereńczak, Krzysztof, Rafał Zapłata, Jarosław Wójcik, Bartłomicj Kraszewski,

Miłosz Mielcarek, Krzysztof Mitelsztedt, Małgorzata Białczak et al., 'ALS-Based Detection of Past Human Activities in the Białowieża Forest – New Evidence of Unknown Remains of Past Agricultural Systems'. *Remote Sensing* 12, no. 16 (18 August 2020): 2657. https://doi.org/10.3390/rs12162657.

Sykes, N. and C. Ameen, 'Preliminary Report on the Animal Bones from Clarendon Palace'. Exeter: University of Exeter, 2021.

Tallis, J. H. and V. R. Switsur, 'Studies on Southern Pennine Peats: VI. A Radiocarbon-Dated Pollen Diagram from Featherbed Moss, Derbyshire'. *The Journal of Ecology* 61, no. 3 (November 1973): 743. https://doi.org/10.2307/2258646.

Taylor, Harvey, *A Claim on the Countryside: A History of the British Outdoor Movement*. Edinburgh: Edinburgh University Press, 2022. https://doi.org/10.1515/9781474473071.

Thompsell, Angela, *Hunting Africa*. London: Palgrave Macmillan, 2015. https://doi.org/10.1057/9781137494436.

Tomlin, Roger, 'A Five-Acre Wood in Roman Kent'. In *Interpreting Roman London: Papers in Memory of Hugh Chapman*, edited by Joanna Bird, Hugh Chapman, M. W. C. Hassall and Harvey Sheldon, 209–15. Oxbow Monograph 58. Oxford: Oxbow Books, 1996.

Tomlin, Roger and Simon Ovin, *Roman London's First Voices: Writing Tablets from the Bloomberg Excavations, 2010–14*. MOLA Monograph 72. London: MOLA (Museum of London Archaeology), 2016.

Tribeca Ltd v. *Joules Limited* (2013).

Trouwborst, Arie, Phillipa C. McCormack and Elvira Martínez Camacho, 'Domestic Cats and Their Impacts on Biodiversity: A Blind Spot in the Application of Nature Conservation Law'. Edited by Juliette Young. *People and Nature* 2, no. 1 (March 2020): 235–50. https://doi.org/10.1002/pan3.10073.

Urban, Christian, Alette A. Blom, Charlotte Avanzi, Kathleen Walker-Meikle, Alaine K. Warren, Katie White-Iribhogbe, Ross Turle et al., 'Ancient *Mycobacterium leprae* Genome Reveals Medieval English Red Squirrels as Animal Leprosy Host'. *Current Biology* 34, no. 10 (May 2024): 2221-2230.e8. https://doi.org/10.1016/j.cub.2024.04.006.

Van Der Plicht, J., L. W. S. W. Amkreutz, M. J. L. Th. Niekus, J. H. M. Peeters and B. I. Smit, 'Surf'n Turf in Doggerland: Dating, Stable Isotopes and Diet of Mesolithic Human Remains from the Southern North Sea'. *Journal of Archaeological Science: Reports* 10 (December 2016): 110–18. https://doi.org/10.1016/j.jasrep.2016.09.008.

Vervust, Soetkin, Tim Kinnaird, Peter Herring and Sam Turner, 'Optically Stimulated Luminescence Profiling and Dating of Earthworks: The Creation and Development of Prehistoric Field Boundaries at Bosigran, Cornwall'. *Antiquity* 94, no. 374 (April 2020): 420–36. https://doi.org/10.15184/aqy.2019.138.

Walker, Thomas, Janet Ridout Sharpe and Hazel Williams, 'Barn Owls and Black Rats from a Rural Roman Villa at Gatehampton, South Oxfordshire'.

Environmental Archaeology 26, no. 5 (3 September 2021): 487–96. https://doi.org/10.1080/14614103.2019.1689805.

Ward, Alastair I., Suzanne Richardson and Joachim Mergeay, 'Reeves' Muntjac Populations Continue to Grow and Spread Across Great Britain and Are Invading Continental Europe'. *European Journal of Wildlife Research* 67, no. 3 (June 2021): 34. https://doi.org/10.1007/s10344-021-01478-2.

Warren, Graeme, Steve Davis, Meriel McClatchie and Rob Sands, 'The Potential Role of Humans in Structuring the Wooded Landscapes of Mesolithic Ireland: A Review of Data and Discussion of Approaches'. *Vegetation History and Archaeobotany* 23, no. 5 (September 2014): 629–46. https://doi.org/10.1007/s00334-013-0417-z.

Webb, Julia C. and Anne E. Goodenough, 'Questioning the Reliability of "Ancient" Woodland Indicators: Resilience to Interruptions and Persistence Following Deforestation'. *Ecological Indicators* 84 (January 2018): 354–63. https://doi.org/10.1016/j.ecolind.2017.09.010.

Whitaker, Katy A., 'Sarsen Stone Quarrying in Southern England'. *Post-Medieval Archaeology* 57, no. 1 (2 January 2023): 143–76. https://doi.org/10.1080/00794236.2023.2173713.

Whitehead, Kenneth G., 'The Muntjac in Britain'. *The Journal of the Bombay Natural History Society* 55 (1958): 158–9.

Wiener, Pamela and Samantha Wilkinson, 'Deciphering the Genetic Basis of Animal Domestication'. *Proceedings of the Royal Society B: Biological Sciences* 278, no. 1722 (7 November 2011): 3161–70. https://doi.org/10.1098/rspb.2011.1376.

Wiessner, Polly W., 'Embers of Society: Firelight Talk among the Ju/'hoansi Bushmen'. *Proceedings of the National Academy of Sciences* 111, no. 39 (30 September 2014): 14027–35. https://doi.org/10.1073/pnas.1404212111.

Williamson, Tom, *An Environmental History of Wildlife in England 1650–1950*. London: Bloomsbury, 2013.

Wordsworth, William, *A guide through the district of the lakes in the north of England: with a description of the scenery, &c., for the use of tourists and residents*. London: Hudson & Nicholson, 1835.

Zu Ermgassen, Sophus O. S. E., Sally Marsh, Kate Ryland, Edward Church, Richard Marsh and Joseph W. Bull, 'Exploring the Ecological Outcomes of Mandatory Biodiversity Net Gain Using Evidence from Early-Adopter Jurisdictions in England'. *Conservation Letters* 14, no. 6 (November 2021): e12820. https://doi.org/10.1111/conl.12820.